Niels Bohr's philosophy of physics

Einstein and Bohr during the Solvay Conference 1930
(By courtesy of the Niels Bohr Library, American Institute of Physics.)

Niels Bohr's philosophy of physics

DUGALD MURDOCH

University of Canterbury

New Zealand

The right of the
University of Cambridge
to print and sell
all manner of books
was granted by
Henry VIII in 1534.
The University has printed
and published continuously
since 1584.

CAMBRIDGE UNIVERSITY PRESS

Cambridge

New York   New Rochelle   Melbourne   Sydney

Published by the Press Syndicate of the University of Cambridge
The Pitt Building, Trumpington Street, Cambridge CB2 1RP
32 East 57th Street, New York, NY 10022, USA
10 Stamford Road, Oakleigh, Melbourne 3166, Australia

First published 1987

Printed in Great Britain at the University Press, Cambridge

British Library cataloguing in publication data
Murdoch, Dugald
Niels Bohr's philosophy of physics.
1. Bohr, Niels   2. Nuclear physics
I. Title
539.7   QC171.2

Library of Congress Cataloguing in publication data
Murdoch, Dugald, 1944–
Niels Bohr's philosophy of physics.
Bibliography:
Includes index.
1. Bohr, Niels Henrick David, 1885–1962.
2. Complimentarity (Physics)   3. Wave-particle
duality.   4. Physics—Philosophy.   I. Title.
QC16.B63M87   1987   530'.092'4   87-11717

ISBN 0 521 33320 2

CONTENTS

QC16
B63
M871
1987
PHYS

To
ANNE-MARIE MURDOCH

PREFACE

What exactly is Niels Bohr's interpretation of quantum physics? And what general philosophical position underlies it? In the following pages I try to present clear and thorough answers to these questions. Bohr's interpretation is undoubtedly of major importance, but there is no universal agreement about what that interpretation is, owing largely to the notorious obscurity of his writings and the fact that his own views are often confused with those of others within the Copenhagen tradition which he founded. My reading is based solely on Bohr's own writings, and it is supported with liberal citations; I defend it against rival readings, but I try to keep polemical discussion to a minimum. When it illuminates Bohr's views, I have made use of material in the Niels Bohr Archive, much of which has now been published in *Niels Bohr: Collected Works* (six vols., eds. Léon Rosenfeld & Erik Rüdinger, North-Holland, Amsterdam, 1972–85). All translations are my own, unless the contrary is stated.

Bohr's interpretation of quantum physics, I believe, contrary to a widely held view, was not the outcome of a ready-made philosophy; rather, it developed gradually from his day-to-day grappling with the problems of the physics. It is best understood, therefore, when seen against the historical background of these problems. My account of this background in the first few chapters is selective, highlighting only those ideas and problems which played a prominent part in the formation of Bohr's views. The historical treatment is not continued beyond the early thirties, since Bohr's views were fully formed by then and changed little thereafter. In relating the relevant physics, I have aimed at simplicity, avoiding undue technicality.

Concerning Bohr's general philosophy of physics I have tried to unravel the realist and non-realist threads which are subtly interwoven in his thinking. The general philosophy which binds these disparate threads together is, I argue, a form of pragmatism. Bohr is not the crude positivist or idealist that he is generally taken to be; indeed, he is much more of a realist. Throughout I contrast Bohr's views with those of Einstein: where the philosophy of physics is concerned, these two thinkers, in my view, tower above their colleagues. Nothing is said about Bohr's attempts to apply his notion of complementarity to subjects other than physics, for I do not believe that they throw light on his philosophy of physics; indeed the contrary is true. I try not only to elucidate Bohr's ideas but also to give a critical appraisal of them.

My own understanding of Bohr's philosophy owes a debt to the writings of other scholars who have sought to explicate Bohr's fascinating but elusive ideas, especially to the writings of Paul Feyerabend, Erhard Scheibe, and C.A. Hooker, which are mentioned in the Notes. On historical matters the works of Max Jammer, Martin Klein, and Klaus Stolzenburg have been helpful. J. Folse's book, *The Philosophy of Niels Bohr: the Framework of Complementarity* (North-Holland, Amsterdam, 1985), came to my notice too late for me to make any use of it. The present book originated in my doctoral dissertation at Oxford University (1980), and I am grateful to Dr. Rom Harré, Prof. Mary Hesse, Dr. William Newton-Smith for their comments on that earlier work. Responsibility for shortcomings in the present work is entirely my own.

I am grateful also to Prof. Aage Bohr for permission to study and make use of the documents in the Niels Bohr Archive, and to Dr. Erik Rüdinger, Director of the Archive, and to the staff of the Archive for the History of Quantum Physics, for assistance with documentary matters.

Christchurch, 1987   D.R.M.

ACKNOWLEDGEMENTS

I am very grateful to the following people and publishers for granting me permission to quote from copyright material: Prof. Birgitte Arrhenius (unpublished letter of O. Klein); Prof. Aage Bohr (published writings of N. Bohr); Elsevier Science Publishers (*Niels Bohr: Collected Works*, ed. L. Rosenfeld, J. Rud Nielsen, and E. Rüdinger); Hebrew University of Jerusalem (A. Einstein, B. Podolsky & N. Rosen, 'Can Quantum-Mechanical Description of Physical Reality be Considered Complete?', *Physical Review*, 47, 1935); Mrs. Elizabeth Heisenberg (published writings of W. Heisenberg); Hermann & Co. (*Albert Einstein–Michele Besso Correspondance*, ed. P. Speziali); MacMillan & Co. (*The Born–Einstein Letters*, ed. M. Born) and (H. Høffding, *A History of Modern Philosophy*, Vol. 2, *Problems of Philosophy* and *Modern Philosophers*); Ox Bow Press (N. Bohr, *Atomic Theory and the Description of Nature*); Munksgaard (H. Høffding, *Correspondance entre Harald Høffding et Emile Meyerson*) Reidel & Co. (J. Mehra, *The Physicist's Conception of Nature*); Prof. P.A. Schilpp (text and figures in *Albert Einstein: Philosopher–Scientist*, ed. P.A. Schilpp); Springer-Verlag (*Wolfgang Pauli: Wissenschaftlicher Briefwechsel*, ed. A. Hermann, K.V. Meyenn, V.F. Weisskopf); Vision Press (*Letters on Wave Mechanics*, ed. K. Przibram).

I

Wave-particle
duality

Niels Bohr's theory of complementarity was an attempt to solve the
enormous problems of interpretation – especially the problem of wave-
particle duality – that beset the quantum theory in the mid-twenties. By 1920
electromagnetic radiation could be conceived of either in terms of the wave
model or in terms of the particle model, though neither conception alone was
wholly adequate to the empirical data. By the mid-twenties the duality
problem was quite general, applying to matter as well as to radiation.
Quantum mechanics, the new quantum theory, could be interpreted
partially in terms of one or other of the two models but not comprehensively
in terms of either.

Bohr's theory can be fully understood only against the background of the
interpretative problems which it was intended to solve; and the root of these
problems goes back to the very origin of the quantum theory, viz. the
quantum hypothesis to the effect that in certain physical processes energy is
transferred discontinuously in discrete amounts. Wave-particle duality was
a development of the continuity-discontinuity duality which the quantum
hypothesis introduced into physics. How did wave-particle duality arise?

1.1
The quantum hypothesis

In 1896 Wilhelm Wien formulated an equation for the energy density of
black-body radiation at temperature T and frequency in the range ν–$\nu + d$:

$$u(\nu, T) = \alpha \nu^3 \exp(-\beta \nu / T),$$

where α and β are constants. This equation agreed with the relevant
empirical data then available, which was confined to radiation of comparatively high frequencies and low temperatures. In 1899 Max Planck put
forward a theoretical derivation of Wien's equation. Within a year, however,
new empirical evidence made it clear that Wien's equation did not hold for

radiation in the low frequency/high temperature range. By skilful mathematical manipulation Planck very quickly suggested a new formula, viz.

$$u(v,T) = Av^3/\exp(Bv/T) - 1$$

which agreed with the data in the low frequency/high temperature range and reduced to Wien's formula in the high frequency/low temperature region.[1] Planck's next task was to provide a theoretical foundation for his new formula.

In his derivation of Wien's equation Planck had made the assumption that a black body may be treated as if it were a collection of linear, harmonic resonators. On the basis of electrodynamics he had shown that the energy density of radiation of frequency v is proportional to the average energy E of a resonator having frequency v and temperature T. He had determined the average energy by way of thermodynamics, expressing the energy of a resonator in terms of its entropy. But when he tried to adopt a similar approach to the derivation of his own equation, he found that the thermodynamic method was unsuccessful. Somewhat at a loss, he turned to the statistical methods of the molecular theory of heat, making use of Ludwig Boltzmann's hypothesis that the entropy of a system in a given state is proportional to the probability of that state. Planck argued that the equation,

$$S_N = k \log W,$$

relates the entropy S_N of a set of N resonators of total energy E_N and frequency v to the quantity W, which was taken to represent the number of ways in which the total energy may be distributed among the N resonators. If, however, E_N were an infinitely divisible quantity, there would be an infinite number of distributions. So, in order that W be determinable by a counting procedure, Planck suggested, following Boltzmann's method, that E_N be treated as finitely divisible, i.e. as consisting of a finite number P of discrete 'energy elements' ϵ (as Planck called them). This made it possible to determine the P energy elements among the N resonators by means of a combinatorial computation. The entropy of a single resonator could then be determined by dividing the equation for S_N by N (since S_N is N-times the entropy of a single resonator). On this basis Planck was able to derive his equation in the form,

$$u(v,T) = (8\pi v^2/c^3) [hv/\exp(hv/kT) - 1],$$

the formula now generally known as 'Planck's energy distribution law'. His derivation was presented at a meeting of the German Physical Society on 14 December 1900.

When presenting his derivation Planck noted that 'the essential point of the whole calculation' is the assumption that the total energy E_N of the resonators is composed of a finite number of equal parts.[2] In adopting this assumption he was following Boltzmann. In his first method of computing the entropy of a gas Boltzmann began by considering a collection of n molecules each of which was restricted to one of a number of discrete energies, $0, \epsilon, 2\epsilon, 3\epsilon, \ldots, p\epsilon$, where p is the total energy of the collection.[3] At an appropriate stage in the calculation Boltzmann took ϵ to be zero. Planck's energy elements, however, unlike Boltzmann's, necessarily have a definite, rather than arbitrary, finite value, owing to Wien's displacement law,

$$B_\nu = \nu^3 F(\nu/T),$$

which requires that the entropy of the resonators is a function of their frequency as well as of their energy. Planck's energy elements necessarily have the fixed value, $\epsilon = h\nu$, where h is the constant β in Wien's energy distribution formula, a constant whose value Planck had already established as 6.55×10^{-27} erg seconds. Planck, then, was unable to take the quantity ϵ to be zero at any point in the calculation. In this way the notion of energy quantisation was introduced into physics.

How did Planck regard his assumption of discrete quanta of energy? In his papers of 1900 and 1901 he gave no indication that he was aware of having made any break with classical physics, let alone having initiated a revolution. The orthodox view on this question is that he held that the energy of individual resonators, but not the energy of the emitted radiation, is quantised, and that the assumption of quantisation was a purely formal device, facilitating computation, but of no, or at least very obscure, physical significance. This interpretation has recently been impugned by T.S. Kuhn, who argues that Planck regarded only the total energy of the collection of resonators as quantised, but not the energy of individual resonators.[4] On this reading the energy and frequency of oscillation of a single resonator could vary continuously within the range $h\nu$. But how, one wonders, could the energy of the whole be quantised but not the energy of the parts over which the whole is distributed? Whichever interpretation is correct, there can be little doubt that Planck intended the assumption of energy quanta to apply only to the resonators and not also to the radiation. The first to suggest that there was something non-classical about Planck's derivation were Albert Einstein and Lord Rayleigh.

Einstein had been drawn to the problem of black-body radiation through his work on the statistical foundations of thermodynamics, upon which he had been engaged since 1902. In his famous paper on light-quanta of 1905 he argued that a classical treatment of the problem leads, not to Planck's law, but to the equation,

$$u(v,T) = (8\pi v^2/c^3)kT,$$

which is now known, somewhat inaccurately, as the Rayleigh–Jeans equation. Einstein noted of course that this equation fails in the high frequency/low temperature range. He noted also that the integral of $u(v,T)$ over all frequencies diverges, and thus the total energy at a given temperature tends towards infinity as the range of frequencies increases – an effect which Paul Ehrenfest later aptly named 'the ultraviolet catastrophe'.[5] Lord Rayleigh, independently of Einstein, arrived at essentially the same finding in his treatment of the problem.[6] Einstein and Rayleigh had employed the Maxwell–Boltzmann equipartition theorem, to the effect that the total kinetic energy of a mechanical system is on average equally distributed among all its degrees of freedom. Planck made no mention of the equipartition theorem in his papers; nor had he employed it. At that time the validity of the theorem was much debated: in 1900 Lord Kelvin regarded it as one of the 'clouds over the dynamical theory of heat and light'.[7] Others, however, supported it. Among them was James Jeans, who criticised Planck for not employing the theorem, and for failing to take ϵ, or equivalently h, to the zero limit, where Planck's equation reduces to the Rayleigh–Jeans formula. Jeans explained away the empirical disconfirmation of the latter formula by arguing that true thermal equilibrium had not been attained in current experiments, and that equipartition was such a slow process that it might take millions of years to be achieved.[8] Jeans, along with Einstein, was the first to stress the profoundly non-classical nature of Planck's law.

In 1906 Einstein maintained that Planck's derivation actually *requires* that the energy of individual resonators is confined to integral multiples of hv, and that the energy changes discontinuously during the emission and absorption of radiation. He went even further, insisting that the energy of the radiation *in vacuo* is also quantised. These assumptions, he stressed, are quite incompatible with classical electrodynamics and the electromagnetic theory of radiation.[9] Moreover, he saw, with unique insight, that the assumption of quantisation must be quite general, affecting the whole of mechanics.

1.2
Einstein's hypothesis of light-quanta

In 1905, when the significance of Planck's radiation law was just beginning to dawn on him, Einstein proposed his hypothesis of the light-quantum. He was well aware of the revolutionary character of the hypothesis, calling it 'very revolutionary' in a letter to his friend Conrad Habicht.[10]

Einstein's hypothesis of the light-quantum grew out of his work in statistical mechanics. He noticed an analogy between black-body radiation and certain characteristics of an ideal gas. He pointed out that if an ideal gas is contained within a vessel of volume V_1 then a change in the volume produces a corresponding change in the entropy of the gas, thus:

$$S_2 - S_1 = N \, k \, \log(V_2/V_1),$$

where S_1 and V_1 are the initial entropy and volume respectively, N is Avogadro's number, and k is Boltzmann's constant. He observed that this equation is formally similar to the one which describes the corresponding change in the entropy of black-body radiation in the domain of Wien's law:

$$S_2 - S_1 = (E/\beta v) \log(V_2/V_1),$$

where E is the energy of the radiation and β a constant. He saw a further similarity. If an ideal gas is contained within a volume V_1, then the probability that all the N molecules of the gas are contained within some subvolume V_2 of V_1 is $W = (V_2/V_1)^N$. Analogously, the probability that all the black-body radiation of frequency v within a volume V_1 is contained within a subvolume V_2 is $W = (V_2/V_1)^{(E/hv)}$. He believed that this similarity was more than merely formal; it reflected a real physical analogy: 'Monochromatic radiation of low density (within the domain of Wien's radiation formula) behaves thermodynamically as if it consisted of energy quanta of magnitude hv'.[11] The energy of such radiation, he suggested, is not distributed continuously over a comparatively wide spatial volume, but discontinuously in a number of independent, localised quanta which move without dividing and which can be emitted and absorbed only as wholes. Einstein called this hypothesis 'a heuristic point of view'.[12]

Einstein believed that the classical electromagnetic theory was correct only for purely optical phenomena (within the low frequency range) such as reflection and diffraction, which have to do with time-averages and not instantaneous values. In processes where instantaneous values become important, as in emission and absorption, the classical theory, couched as it is in terms of continuous spatial functions, may break down.[13] Such a breakdown occurs, Einstein suggested, in the photo-electric effect, and in the phenomena of photo-luminescence and photo-ionisation.

In the photo-electric effect, for example, the kinetic energy of electrons emitted by a metal illuminated *in vacuo* with ultraviolet light is independent of the intensity of the incident light but proportional to its frequency. In terms of the wave model of radiation one would expect the energy of the emitted electrons to be proportional to the intensity of the light, and the intervals between the onset of the illumination and the emission of electrons to be longer than it is found to be. Philipp Lenard explained this effect in terms of the wave model as a resonance phenomenon: electrons vibrating in resonance with radiation of the same frequency are ejected with an energy which is proportional to the vibration frequency; the incident radiation thus functions merely as a trigger for the release of electrons, which already possess within the atom the energy required for their release. This explanation was generally accepted until 1912.[14]

Einstein explained the photo-electric effect in terms of his light-quantum hypothesis, according to which individual light-quanta transmit kinetic energy to individual electrons, thus ejecting them from the atoms of the metal. Moreover, he predicted that the maximum kinetic energy of the ejected electrons is a linear function of the frequency of the light – a bold prediction that was not confirmed until ten years later.

It is characteristic of Einstein that his main reasons for questioning the general validity of the electromagnetic theory of radiation were of a peculiarly fundamental sort. His work on the special theory of relativity had convinced him that the ether, which seemed to be presupposed by the electromagnetic theory, did not exist. Thus, if there was no ether, radiation need not be conceived of as a wave entity. Moreover, he was troubled by what he saw as a conceptual disparity, a profound formal difference, between the theoretical concepts of the classical mechanical theory of matter and James Clerk Maxwell's theory of radiation.[15] Whereas the energy of matter is concentrated in discrete entities, particles, the energy of radiation is diffusely spread over a comparatively wide spatial volume; while a particle is a localised point-like mass, a wave is a periodic disturbance in a continuous, spatially extended region. Moreover, Einstein was troubled by an asymmetry in the interaction between radiation and matter: whereas emission of radiation could be conceived of as the production of a spherical wave by a charged body, the absorption process could not be conceived of as the inverse of the former process.[16] In 1905 he thought that these difficulties required a modification of the electromagnetic theory that would make room for light-quanta.

1.3
Wave-particle duality, 1905–10

In the following year, 1906, Einstein argued that Planck's derivation of his radiation law was inconsistent: he had combined two expressions for the average energy of a resonator: on the one hand the expression,

$$E(v,T) = u(v,T) \ (8\pi v^2/c^3),$$

and on the other hand the expression,

$$E(v,T) = hv/\exp(hv/kT) - 1.$$

Whereas the former expression is derived from classical electrodynamics and presupposes that the energy of a resonator is a continuously variable quantity, the latter expression is derived from statistical mechanics and presupposes that the energy varies discontinuously in discrete amounts $\epsilon = nhv$. If the energy of a resonator varies discontinuously, then classical electrodynamics cannot legitimately be used for the calculation of the average energy of a resonator. If one were to employ classical electrodynamics, one would have to assume that, although the classical theory is strictly speaking incorrect, it happens as a matter of fact to yield the correct values for the average energy of the resonator – an assumption which would be impossible if the amounts $\epsilon = nhv$ by which a resonator gains or loses energy were in general small in relation to the average energy of the resonator; but this is not the case for high frequencies and low temperatures. Einstein saw that the assumption of discontinuity, that energy is quantised in discrete amounts, is *necessary* for the derivation of Planck's radiation law; indeed, he remarks that the hypothesis of light-quanta is implicit in Planck's derivation.[17]

The fact that Planck's derivation is based upon the incompatible assumptions of continuity and discontinuity shows, Einstein held, not that Planck's law is not universally valid, but rather that one of the theories from which it is derived, viz. classical electromagnetic theory, requires major modification. Again, his special theory of relativity gave him other grounds for suspecting the electromagnetic theory.

Up until 1909 very few physicists accepted the quantum hypothesis; these few included Johannes Stark, Wien, Ehrenfest, Max von Laue, H.A. Lorentz, and Planck. Lorentz was a recent convert: between 1903 and 1909 he had struggled in vain to provide a classical basis for Planck's law.[18] The quantum hypothesis began to receive a wider recognition only after the first Solvay Conference at Brussels in 1911. This conference brought home to many physicists the major significance of the hypothesis. Why in fact did the hypothesis come to be generally accepted? Firstly, because of the persistent failure of Planck and others to derive Planck's empirically successful formula from classical assumptions alone. Secondly, because of its heuristic power: for example, Planck was able to calculate from his theory the values of Avogadro's number, the mass of the hydrogen atom and the charge of the electron; and in 1910 Walter Nernst confirmed Einstein's prediction (based on the quantum hypothesis) that the specific heats of liquids and solids should approach zero at very low temperatures.

Although by 1912 the quantum hypothesis was widely accepted, there was no universal agreement about its scope. Most physicists shared Planck's view that the hypothesis applied only to matter, on the grounds that it was incompatible with Clerk Maxwell's equations. The quantum of action was, Planck held, a constant of matter restricting the absorption and emission of radiation; once radiated, energy propagates through space in spherical waves.[19] Planck could legitimately hold this view, since the thesis that the energy of harmonic resonators is quantised does not logically entail that the radiation is emitted and absorbed in the form of light-quanta. I do not believe that Einstein thought there was a logical entailment here; he had other grounds for proposing the hypothesis of light-quanta. For example, a spherical wave, spread over an enormous distance, would have to deliver its quantum of energy practically instantaneously when it is absorbed by an atom; if the quantum of energy were spread throughout the entire spherical wave-front, the instantaneous transmission of energy would be at odds with the special theory of relativity. Moreover, Einstein argued, light-quanta are a necessary consequence of the energy fluctuations of radiation *in vacuo*.

In a paper of 1909 Einstein argued that the mean-square energy fluctuations of radiation in a small volume V of an isothermal enclosure is the sum of two terms:

$$\langle \epsilon^2 \rangle = E(v,T)hv + [E(v,T)^2 c^3/8\pi v^2]V dv,$$

the former explicable on the assumption that the radiation is composed of mutually independent quanta of energy $\epsilon = hv$; and the latter explicable as the expression for an interference between partial waves.[20] The energy fluctuations seemed to have two independent causes, the one corpuscular, predominating in the high-frequency range (the Wien limit of Planck's law), the other undulatory, predominating in the low-frequency range (the Rayleigh–Jeans limit).

Einstein was struck by the dimensional equivalence between Planck's constant h and the constant e^2/c ('e' signifying the electric charge, 'c' signifying the light-velocity constant). He held that the discrete electric charge e is as foreign to electrodynamics (which permits a continuously variable charge) as the quantum of action h is foreign to electromagnetics. What was required was a modification of these classical theories that entailed the quantum structure of both electricity and radiation, i.e. a field theory of matter and radiation which entailed the existence of both the electron (the quantum of electricity) and the light-quantum (the quantum of radiation); such a theory, he suggested, would employ non-linear equations.[21]

It was clear to Einstein that the light-quantum hypothesis could not simply supersede the wave theory: what was needed was a theory uniting the continuous and the discontinuous aspects of radiation, a 'fusion of the wave and the emission theories'. He suggested that the electromagnetic field is bound to singular points (light-quanta) just as the electrostatic field is bound to electrons. On this view the entire energy of the electromagnetic field would be concentrated in these localised singularities, which would be the centres of force fields having the character of waves. If many such singularities were present in a small volume, the corresponding fields would be superimposed, forming an undulatory field similar to the classical electromagnetic field. His suggestion was intended to show that the two characteristics of radiation, the continuous and the discontinuous (both implicit in Planck's formula $\epsilon = hv$) need not be regarded as mutually incompatible.[22]

Einstein put forward these ideas at the Conference of German Scientists at Salzburg in September 1909. In the discussion following Einstein's paper Planck argued that the light-quantum hypothesis necessitated the abandonment of Clerk Maxwell's equations – a step he did not think was yet warranted, even for high-frequency radiation – and that interference phenomena presented a major difficulty for the hypothesis. The quantum of action, he thought, had to do only with the interaction between radiation and matter, and not with radiation *in vacuo*. Einstein replied that interference phenomena could not be explained on the assumption that light-quanta were mutually independent; they could be explained, however, on the assumption that light-quanta interact through their associated fields. It was the difficulty of explaining interference phenomena that convinced him that light-quanta could not be mutually independent, as he had formerly thought.[23] This reply proved unsuccessful, for that same year G.I. Taylor showed that interference effects can be built up comparatively slowly by single light-quanta, the implication being that the interference pattern cannot be due to the mutual interaction between individual light-quanta.

Einstein's change of mind regarding the independence of light-quanta was due to Lorentz's criticism of the hypothesis. Lorentz published his objections in an influential paper of 1910. He argued that interference and diffraction effects cannot be explained on the assumption that light-quanta are mutually independent, indivisible, point-like entities. He pointed out that widely separated beams of monochromatic radiation can produce interference effects; if such beams consist of light-quanta, then each light-quantum would have to be very long in the direction of its propagation (over 80 cm in some cases). Again, the image of a star observed through a telescope increases in size if the mouth of the telescope is narrowed. This spreading of the image can be explained only if light-quanta are extended over a wide front in the direction of their propagation; in which case they cannot be point-like entities.[24]

Between 1909 and 1918, as his letters testify, Einstein spent an enormous amount of time trying to construct a comprehensive theory of light-quanta that accounted for interference effects, but, as he admitted, without success.[25]

1.4
Wave–particle duality, 1911–22

Einstein's light-quantum hypothesis met with almost universal rejection, Stark (in 1909) being, it seems, his sole ally. Einstein's explanations of the photo-electric, photo-luminescence, and photo-ionisation effects were generally regarded as ingenious but unconvincing. There can be little doubt, moreover, that the discovery by Walter Friedrich and Paul Knipping in 1912 of the diffraction of X-rays (predicted by von Laue) made it even more difficult for Einstein's contemporaries to accept the light-quantum hypothesis: even high-frequency radiation, concerning which Einstein's hypothesis was at its most plausible, appeared to be a wave phenomenon.

The situation was perplexing, however, for in 1912 experimental evidence supporting Einstein's photo-electric equation was forthcoming in the work of A.L. Hughes, and of O.W. Richardson and K.T. Compton.[26] Much stronger support was provided in 1915 by R.A. Millikan, who showed that the kinetic energy of the emitted electrons does not exceed a maximum value for incident radiation of a given frequency, and that the energy of the electrons is a linear function of the frequency of the incident light, exactly as Einstein had predicted in 1905.[27] In spite of this striking confirmation of Einstein's equation, Millikan did not regard his experiments as confirming the light-quantum hypothesis, which he held to be wholly untenable; in this case, he thought, erroneous theory had led to the discovery of empirical relations of the greatest importance. Like most of his contemporaries Millikan believed the photo-electric effect to be a resonance phenomenon.[28]

The following year (1916) Einstein took his light-quantum hypothesis a stage further when he formulated the probability laws governing the emission and absorption of radiation by an atom. Basing his theory on Bohr's notion of the stationary states of the atom, he was able to derive Bohr's frequency condition, $E_1 - E_2 = h\nu$, and Planck's radiation law 'in an astonishingly simple and general way'. He went on to argue that when an atom absorbs or emits radiation of energy $h\nu$ its momentum is changed by an amount $h\nu/c$ in the direction of the incident radiation in the case of absorption, in the direction opposite to that of the emitted radiation in the case of 'induced' emission, and in a direction determined by chance in the case of 'spontaneous' emission (as Bohr was to call it).[29] This was a marvellous theoretical achievement.

According to classical electrodynamics emission of radiation from an atom should occur in the form of spherical waves; hence there should be no change in the momentum of the emitting atom. Einstein's hypothesis contradicts this: 'Outgoing radiation in the form of spherical waves does not exist'.[30] He considered the hypothesis of momentum change the main result of his paper.[31] The weakness of the theory in his view was that although it brings us closer to a complete quantum theory of radiation, it brings us no closer to the wave theory; and moreover, it leaves the time of occurrence and direction of some elementary processes to chance. He was confident, however, that the hypothesis was correct and also that some underlying causal mechanism governs the 'spontaneous' emission processes.

Now that light-quanta were endowed with momentum, they began to look more and more like particles; they were no longer just indivisible, point-like entities. One wonders why it took Einstein so long to ascribe a momentum to the light-quantum.

Describing his findings in a letter to Michele Besso, Einstein remarked, 'Thus the light-quanta are as good as proved.'[32] As the years went by Einstein attempted to disprove the wave theory. In 1921, for example, he thought he had found a crucial experiment: he argued that according to the wave theory a high-temperature radiation field should produce a noticeable Stark effect on the emitting and absorbing atoms, whereas according to the light-quantum hypothesis the Stark effect should occur only on a few atoms – a difference which ought to be detectable. Nothing seems to have come of this suggestion.[33] Later that year he argued that, whereas according to the wave theory the frequency of light emitted from a moving atom should vary with the angle of observation, according to the light-quantum hypothesis the frequency should be uniform. Walter Bothe and Hans Geiger performed this experiment and showed that the frequency was uniform.[34] Einstein took this result to show that the electromagnetic field does not exist, and that emission of radiation is an instantaneous process – a result which he regarded as his most impressive scientific experience in years.[35] Ehrenfest, however, objected that the experiment showed no such thing, since Einstein had failed to take into account the fact that the emitted light has the form of a wave-packet and not of an infinite wave-train, which implies that no shift in frequency should occur. Einstein called his error 'a monumental blunder'.[36]

Although Einstein failed in his attempt to disprove the wave theory, the Compton effect soon provided a further, brilliant confirmation of the light-quantum hypothesis.

1.5
The Compton effect

It had been shown by D.C.H. Florance in 1910 and by J.A. Gray in 1913 that the secondary radiation produced by irradiating a thin metal plate with gamma rays is softer (i.e. less penetrating) than the primary radiation, and that the greater the angle of scatter the softer the secondary radiation. When A.H. Compton investigated this effect in 1921 he came to the conclusion that the secondary radiation is of a different wavelength from the primary radiation. He held that the soft secondary radiation is not scattered radiation but a kind of fluorescent radiation emitted by beta rays (fast electrons emitted by the primary beam). The fast beta particles cause a Doppler shift in the wavelength of the fluorescent radiation, which accounts for the directional variability of the degree of softening. In 1921 he showed by experiment that 'scattered' monochromatic X-rays undergo a change in quality and frequency, a result which he took to confirm his interpretation of the soft secondary radiation as fluorescent.[37]

In March 1922 Compton discovered the total internal reflection of X-rays – an important finding, which strengthened his belief in the wave theory of radiation. In October 1922 he again argued in favour of his explanation of the softened secondary radiation as fluorescent, though he suggested, as an afterthought, that a different explanation could be given in terms of the quantum theory. If each scattering electron receives a whole quantum and re-radiates it in a definite direction, then the momentum of the scattering electron will be increased by an amount $h\nu/c$, which will produce a Doppler shift as the scattered radiation is observed at different angles. Nevertheless, he thought that the existence of interference phenomena made the quantum-theoretical explanation very doubtful.[38]

Compton soon came to see, however, that if the observed change of wavelength of the gamma rays scattered at an angle of 90 degrees was due to a Doppler shift, the velocity of the beta particles would have to be about half the velocity of light *in vacuo*; and only a few (if any) beta particles could possibly be moving at such a high speed. It was the implausibility of this consequence that led him to re-consider the explanation in terms of light-quanta. The change in wavelength could be explained on the assumption that light-quanta of the incident radiation of energy $E = h\nu$ and momentum $p = h/\lambda$ collide with individual electrons in the scattering substance and are deflected with reduced energy $E' = h\nu'$ and momentum $p' = h/\lambda'$, which are conserved in the recoil energy and momenta of the deflecting electrons, the wavelengths of the scattered light-quanta being increased by the amount $\lambda'' = \lambda - \lambda' = h/p - h/p'$. Compton presented this explanation at a meeting of the American Physical Society in Chicago at the beginning of December 1922.[39] As R.H. Stuewer has shown, Compton, committed as he was to classical electrodynamics, tenaciously strove to find an explanation in classical terms right up until the eleventh hour.[40]

The following year (1923) the cloud-chamber experiments of C.T.R. Wilson and Bothe provided evidence of the tracks of the recoil electrons, the length and direction of the tracks agreeing with the light-quantum theory.[41] Wilson pointed out, however, that the cloud-chamber experiments did not provide conclusive evidence for the existence of light-quanta, since the evidence was insufficient to determine whether a light-quantum scattered by an electron is emitted in a specific direction or over a continuous wave-front. Wilson believed that the radiation is scattered in the form of spherical waves. Compton rejoined, however, that according to the classical wave theory the effective velocity of the recoil electrons necessary for the required Doppler shift would make the length of their tracks sixteen times shorter than the length calculated according to the light-quantum theory, which is the length actually observed in the cloud-chamber experiments.[42]

Compton's explanation of his experiments in terms of the light-quantum theory was by no means immediately accepted by the majority of physicists; indeed, there was considerable opposition to it. Several distinguished physicists tried to explain the Compton effect in terms of a more or less modified classical electrodynamics.[43] The staunchest opponent of the explanation in terms of light-quanta was, as we shall see below, Niels Bohr. This strong opposition notwithstanding, the Compton effect and its explanation in terms of quanta greatly enhanced the status of Einstein's light-quantum hypothesis.

In November 1922 Einstein was awarded the Nobel Prize for Physics (for 1921) – 'for his services to theoretical physics and especially for his discovery of the law of the photo-electric effect'. Why did it take so long – from 1905 to 1922 – for the light-quantum hypothesis to be taken really seriously? Firstly, because the wave model of radiation was firmly entrenched: diffraction and interference phenomena, it was generally thought, conclusively ruled out the light-quantum model. Secondly, phenomena which the wave theory could explain only with enormous strain, such as the photo-electric effect, were regarded merely as anomalies for the wave theory, problems which would be solved once the mechanism of the interaction between radiation and matter was better understood. As time went by, however, attempts to suggest a convincing mechanism failed, while the explanatory and predictive power of the light-quantum hypothesis steadily increased. This remarkable historical development nicely illustrates Pierre Duhem's thesis that an established theory cannot be killed off at a stroke. The light-quantum hypothesis came to be taken really seriously only after it had clearly become empirically progressive (as Imre Lakatos would have put it).

By 1923, then, the wave-particle duality of radiation could no longer be ignored: it was firmly established as one of the most perplexing and recalcitrant problems of physics. There were two incompatible models of radiation, each well confirmed; yet evidence for the one constituted counter-evidence for the other.

Niels Bohr
and wave-particle duality

What was Bohr's response to the problem of wave-particle duality? Before considering this question in detail it will be useful to survey briefly Bohr's work in physics between 1911 and 1923.

2.1
Bohr and the 'old' quantum theory

In 1911 Bohr completed his doctoral dissertation on the classical electron theory of metals. Developed by Sir J.J. Thomson and Lorentz, the theory explained electrical, magnetic and thermal properties of metals in terms of the motions of free electrons. Although Bohr made no use of the quantum hypothesis in his dissertation, he noted that classical electrodynamics was incapable of explaining Planck's radiation law and entailed the ultraviolet catastrophe.[1]

After completing his dissertation Bohr pursued his post-graduate studies under Thomson at Cambridge University. Finding that Thomson was little interested in his work, Bohr went on to Manchester University in March 1912, to work under Ernest Rutherford. Bohr's move to Manchester was very opportune: the quantum hypothesis was just beginning to receive a wider recognition, thanks largely to the first Solvay Conference in 1911, which Rutherford attended; Rutherford's laboratory, moreover, was one of the most active centres in the investigation of radio-activity.

In order to explain the results of experiments on the passage of alpha and beta particles through matter Rutherford had revived Hantaro Nagaoka's (1904) planetary model of the atom. On this model an atom consists of a positively charged nucleus which is orbited by negatively charged electrons, the number of which corresponds to the number of charges on the nucleus. While occupying a minute part of the volume of the atom, the nucleus makes up almost all of its mass. On this model, however, the atom has an unstable structure: a single negatively charged particle revolving round a positively charged nucleus would radiate energy continuously and rapidly spiral into the nucleus; moreover, a multi-electron atom would be mechanically unstable. The problem of the electrodynamic and mechanical stability of the Rutherford atom was ready and waiting for Bohr on his arrival in Manchester. He quickly tackled the problem and achieved a brilliant solution which explained not only the stability of the hydrogen atom but its line spectra as well. His theory, like Einstein's theory of the light-quantum, was of astonishing novelty. In a series of three remarkable papers published in 1913 Bohr presented a quantum theory of atomic structure. In the first paper he proposed a model which explained the line spectra of the hydrogen atom; in the second paper he suggested how his theory might be developed to explain the structure of multi-electron atoms and to account for the periodic system of the elements; and in the third he suggested a theory of the structure of molecules.[2]

Bohr's theory was based upon the following hypotheses:

(*a*) Atoms possess a finite number of discrete energy states ('stationary states'); when in one of these states an atom can neither emit nor absorb radiation.

(*b*) An atom emits or absorbs energy discontinuously, in the form of monochromatic radiation, when it passes from one stationary state to another, the frequency of the emitted or absorbed radiation being proportional to the difference between the energies of the stationary states in question: thus, $\nu = E/h = (E_1 - E_2)/h$.

(*c*) The ratio between the kinetic energy of an electron and its frequency of revolution ω in a stationary state is $E = nh\omega/2$; in the case of a circular orbit the angular momentum of the electron is an integral multiple of $h/2\pi$.

(*d*) The dynamical equilibrium of the atom in a stationary state is governed by the laws of classical mechanics, though these laws fail to govern transitions between stationary states.

These hypotheses were manifestly contrary to classical physics: classical electrodynamics was flatly contradicted, and practically the only role left for classical mechanics was to account for certain average values of the motions of the electrons. Bohr's theory brought the notion of discontinuity, implicit in the quantum hypothesis, into particularly sharp focus: the discontinuous transitions between stationary states came to be known colloquially as 'quantum jumps', a vivid metaphor for a kind of event that lay beyond the ken of classical physics.

Bohr's theory was highly successful: not only did it explain Balmer's formula governing the line spectrum of hydrogen, but it also accounted for Rydberg's formula describing the spectra of the heavier elements, and it yielded a value for Rydberg's constant which agreed well with the value obtained by experiment. Moreover, it made several bold and novel predictions: it suggested that the Pickering series of lines belonged not to hydrogen, as was generally believed, but to helium – a surprising attribution which was soon confirmed by experiment. When Einstein was told of this result he is reported to have proclaimed, '. . . this is an enormous achievement. The theory of Bohr must then be right'.[3] Furthermore, the theory predicted the Lyman, Brackett, and Pfund series, which were observed in 1914, 1922, and 1924 respectively. It also provided an explanation of the important Frank–Herz experiment of 1914, since it entailed that in a collision between free electrons and atoms the free electrons lose energy in discrete amounts – the free electrons cannot lose an amount of energy which is less than the difference between the ground state and the first 'excited' state. In his 'Autobiographical Notes' Einstein writes:

> That this insecure and contradictory foundation [i.e. Planck's quantum theory] was sufficient to enable a man of Bohr's unique instinct and tact to discover the major laws of the spectral lines and of the electron shells of the atoms appeared to me like a miracle – and appears to me as a miracle even today.[4]

Despite its weird postulates Bohr's theory was quickly recognised to be a major breakthrough in the development of the quantum theory. This nicely illustrates the point, stressed by Sir Karl Popper, that a novel theory in physics is accepted only if it is thought to have empirical content in excess of that of the prevailing theories. In 1915 Arnold Sommerfeld generalised the theory by considering several degrees of freedom and treating the motion of the electron in terms of the special theory of relativity. This elaboration enabled Sommerfeld to explain the fine structure of the line spectrum of hydrogen, and it enabled his pupils, Paul Epstein and Peter Debye, to explain the Stark effect and the Zeeman effect.

Bohr's theory initiated the quantum theory of atomic structure, and laid down a programme of research which was to occupy him and many others for the next ten years. Between 1915 and 1921 the Bohr–Sommerfeld theory developed steadily. Bohr's two most important contributions to this development were his correspondence principle – to the effect that in the limit of high quantum numbers the results of the quantum theory should coincide with those of the classical theories – and his (1921) theory of the periodic system of the elements. The latter theory was based on the hypothesis that atoms are built up by the successive binding of electrons in orbits around the nucleus, in such a way that the binding of an electron to an atom does not change the quantum numbers of the pre-existing stationary states of the atom in question. This theory was remarkably successful: for example, it led to the discovery of element 72 (hafnium, named after the Latinised name for Copenhagen), the properties of which it had correctly predicted.

By 1922, however, the Bohr programme was running out of steam. Although it had many successes to its credit, intractable problems remained, such as the problem of determining the energy levels of the helium atom, and the anomalous Zeeman effect. Problems such as these were not effectively solved until the advent of quantum mechanics in 1925.

During the period 1913–23 Bohr's work was confined largely to the quantum theory of atomic structure, and he was little concerned with the quantum theory of radiation as such. The emission and absorption spectra, and the X-ray spectra of the elements did of course provide the main empirical evidence for testing the atomic theories, but the nature of the process responsible for spectral radiation was not a question of paramount interest for Bohr. Although he held that the atom emits and absorbs radiation in discrete quanta of energy, he maintained (with Planck) that the radiated energy is transmitted through space in continuous wave fronts, and that it is not necessary to assume (with Einstein) that discrete emission and absorption imply discrete propagation of radiation *in vacuo*. The main weakness in classical physics, he thought, lay not with Clerk Maxwell's equations, which he believed to be valid for empty space, but with mechanics.[5]

As the evidence for the light-quantum theory increased, however, it came to seem more and more problematical that the propagation of radiation *in vacuo* should be continuous while its emission and absorption is discontinuous. How exactly did Bohr respond to this situation, the problem of wave-particle duality?

2.2
Bohr's attitude to the light-quantum hypothesis

In the following discussion I shall use the term 'model' to denote a physical interpretation of the mathematical formalism of a physical theory which includes a pictorial representation or visualisable conception of the objects to which the theory is intended to be applied. Bohr's usual term for a model in this sense is 'picture'. I shall call a model 'realistic' if it purports to provide a true representation of the objects of the theory in question. Realistic models are seldom complete representations: they generally bear certain similarities and certain dissimilarities to the modelled object; these Mary Hesse aptly calls the 'positive analogy' and the 'negative analogy' respectively.[6] I shall call a model 'formal' if it is not regarded as providing a true representation of the objects of the theory in question.

After 1923 one of the main problems in the interpretation of the quantum theory as Bohr saw it was how to reconcile the idea of the discontinuous exchange of energy between the atom and the radiation field with the classical notion of the continuous transmission of radiation *in vacuo*. Rather than adopting Einstein's light-quantum hypothesis, Bohr sought a way to render these two notions mutually consistent.

Unlike Einstein, he regarded the disjunction between the wave hypothesis and the light-quantum hypothesis as irreconcilably mutually exclusive. Moreover, like most physicists he was very reluctant to accept the light-quantum hypothesis, for he believed that only the wave hypothesis could provide a satisfactory explanation of interference phenomena. In a lecture of 1920 he says that such phenomena

> . . . cannot possibly be understood on the basis of a theory such as that of Newton . . . In fact, the picture provided by Einstein's conceptions looks very much like Newton's, and it can no more than that give any sort of explanation of the interference phenomena.[7]

In his Nobel Prize Address of 1922 he writes:

> In spite of its heuristic value, however, the hypothesis of light-quanta, which is quite irreconcilable with so-called interference phenomena, is not able to throw light on the nature of radiation. I need only recall that these interference phenomena constitute our only means of investigating the properties of radiation and therefore of assigning any closer meaning to the frequency which in Einstein's theory fixes the magnitude of the light-quantum.[8]

Bohr's point here is that the values of the physical quantities ν and λ are determined by experiments involving diffraction or interference effects, which can be understood only in terms of the wave theory: it is only in terms of the wave theory that the notions of frequency and wavelength have any well-defined physical meaning.[9] He regards the light-quantum model as having some heuristic, but no realistic, significance; it simply illustrates the

difficulty of reconciling the discontinuity inherent in atomic processes with
the continuity presupposed by classical electrodynamics. He continues
almost immediately:

> The hypothesis of light-quanta, therefore, is not suitable for giving a picture of
> the processes, in which the whole of the phenomena can be arranged . . . The
> satisfactory manner in which the hypothesis reproduces certain aspects of the
> phenomena is rather suited for supporting the view . . . that, in contrast to the
> description of natural phenomena in classical physics in which it is always a
> question only of statistical results of a great number of individual processes, a
> description of atomic processes in terms of space and time cannot be carried
> through in a manner free from contradiction by the use of conceptions
> borrowed from classical electrodynamics . . .[10]

The partial viability of the light-quantum hypothesis shows only that there
can be no single, coherent spatio-temporal model of the interaction between
radiation and matter.

It is clear that Lorentz's paper of 1910 on the light-quantum had a
profound influence on Bohr's thinking.[11] Oskar Klein, who was Bohr's
assistant at the time, notes that Bohr frequently stressed Lorentz's
arguments for the incompatibility between the light-quantum hypothesis
and facts concerning the interference of light waves.[12] At this time (1922–23)
Bohr regarded the quantum theory as purely formal, in the sense that its
mathematical equations are unamenable to interpretation in terms of
realistic models; the theory says nothing about the real nature of atomic
processes, providing merely a mathematical description of the statistical laws
governing them. He did think, however, that the electromagnetic theory had
realistic import:

> Our whole knowledge of the nature of radiation . . . rests solely on these
> phenomena, in the closer consideration of which the formal nature of the
> quantum theory stands out particularly clearly.[13]

Until 1925 Bohr was not concerned so much with the issue of wave-particle
duality as with the continuity-discontinuity duality. The main problem as he
saw it was the apparent inconsistency, or at least conceptual disharmony,
between the quantum theory, which implies discontinuous emission and
absorption of radiation, and the electromagnetic theory, which implies
continuous emission and absorption. In order to reconcile the discontinuity
of the one with the continuity of the other what was required was, not a
replacement of the electromagnetic theory, but a radical departure from
classical mechanics.

2.3
Bohr's attitude to the Compton effect

Compton's explanation of his experimental findings in terms of the light-quantum hypothesis aggravated the difficulty of reconciling the notion of the continuous propagation of radiation *in vacuo* with the notion of the discontinuous interaction between radiation and matter.

Bohr was in the USA during the autumn of 1923. Curiously enough he makes no mention of Compton's work in his Silliman lectures at Yale. It is clear, however, that he was still not prepared to modify his assessment of the light-quantum hypothesis, regarding which he writes:

> This cannot however be considered as a serious theory of light transmission. Light is not only a flow of energy, but our description of radiation involves a large amount of physical experience involving optical apparatus including our eyes for the understanding of the working of which nothing seems satisfactory except the wave theory of light. No significance for the quantity v without waves.[14]

Bohr certainly knew of Compton's work. He had been informed about it early in 1923 by Sommerfeld, a keen supporter of Compton, who wrote Bohr saying that he thought Compton's work falsified the wave theory of X-rays.[15] Bohr was present at the meeting of the American Physical Society in Chicago between 30 November and 1 December 1923, at which he read a paper entitled 'The Quantum Theory of Atoms with several Electrons'. Compton relates that in conversation Bohr and C.G. Darwin had expressed a preference for the wave model and an abandonment of the laws of conservation of energy and momentum. Bohr totally rejected the explanation of the Compton effect in terms of the light-quantum hypothesis. By 1923 he was already working out his own solution to the radiation problem, one involving the abandonment of the classical conservation laws.

Bohr's opposition to the light-quantum hypothesis as late as 1923 was by no means unique; nevertheless it was of extraordinary pertinacity. Werner Heisenberg recalls that at this time Bohr was the only physicist he knew who insisted that the light-quantum hypothesis was false: Bohr once told him, facetiously no doubt, that if Einstein were to send him a telegram announcing irrefragable evidence for the existence of light-quanta, the telegram could reach him only in virtue of the waves that were there.[16]

Bohr's and Einstein's conceptions of the radiation problem at this time were diametrically opposed. Einstein believed that the main source of the weaknesses in classical physics lay in the electromagnetic-electrodynamic theory; Bohr believed that the main source lay in mechanics. Whereas Einstein held that the electromagnetic theory was correct only if interpreted in a statistical sense, Bohr held that the conservation laws were correct only in a statistical sense. While for Einstein the wave model was merely a pictorial metaphor for the laws governing the motions of photons, for Bohr the photon model was a pictorial metaphor for the laws governing the interactions between radiation and individual atoms.

In 1921 Bohr came to think that what was required for the solution of the radiation problem was the discovery of some, as yet unknown, non-classical mechanism governing the discontinuous interaction between matter and radiation, a mechanism which would be incompatible with the strict conservation of energy and momentum.[17] Darwin had suggested in 1919 that the incompatibility between the electromagnetic theory and the quantum theory would be removed if energy were conserved only on the average taken over many individual processes. He noted that Einstein had tried this approach, unsuccessfully.[18] Einstein had indeed tried it, in 1910, but quickly abandoned it.[19] Darwin developed a quantum theory of the dispersion of radiation based upon the idea of a statistical conservation of energy. He argued that an illuminated atom has a certain probability of emitting a spherical wave-train, the probability being determined by the intensity of the incident light.[20] Although Bohr was critical of Darwin's theory, pointing out *inter alia* that it could not account for the dispersion of light of very weak intensity, he was nevertheless sympathetic to the idea of a statistical conservation of energy and momentum.[21] Towards the end of 1923 he found a way of implementing this idea.

2.4

The Bohr–Kramers–Slater theory

In an attempt to reconcile the continuity implicit in the classical theory with the discontinuity presupposed by the quantum theory, J.C. Slater suggested that the atom in a stationary state be regarded as a set of virtual harmonic oscillators (i.e. resonators) whose frequencies are those of possible transitions between stationary states, and that atoms communicate with one another by means of virtual radiation fields produced by the oscillators. The idea of virtual radiation fields was intended to reconcile the light-quantum hypothesis with classical electromagnetics. Thus, the intensities of the virtual fields emitted and absorbed by an atom in a stationary state determine the probabilities of emission and absorption of light-quanta or photons (as they came later to be called), the emission of a photon by one atom being coupled uniquely with its absorption by another.[22]

Slater's notion of virtual radiation, i.e. unobservable radiation which transmits no energy, was the sort of coupling mechanism that Bohr had been looking for. A year previously he had written of the coupling mechanism:

> . . . it must be assumed that this mechanism . . . becomes active when the atom is illuminated in such a way that the total reaction of a number of atoms is the same as that of a number of harmonic oscillators in the classical theory, the frequencies of which are equal to those of the radiation emitted by the atom in the possible processes of transition, and the relative number of which is determined by the probability of occurrence of such processes of transition under the influence of illumination.[23]

Bohr even speaks of 'latent reactions of radiation, which answer to the harmonic components corresponding to the respective processes of transition'.[24] This is virtually the inchoate idea of virtual radiation that Slater made explicit.

The idea of virtual radiation suggested a mechanism that would determine the probabilities of transition between the stationary states of the atom, in such a way that the energy and momentum exchanged between individual atoms and radiation would not be strictly conserved in individual interactions, but only on the average over numerous transitions. Bohr discarded Slater's notion that the virtual radiation field determines the emission and absorption of photons. Bohr's assistant, H.A. Kramers, had argued in correspondence with Slater that a greater independence, rather than a close coupling, between transition processes in widely separated atoms would be expected on the basis of Slater's hypothesis.[25] Bohr simply dropped Slater's use of the photon hypothesis and added the notion of statistical conservation of energy and momentum. In a letter to Slater he writes:

> It was just the completion which your suggestion of radiative activity of higher quantum states apparently lent to the general views of the quantum theory with which I had been struggling for years, which made me welcome your suggestion so heartily. Especially I felt it was far more harmonious from the point of view of the correspondence principle to connect the spontaneous radiation with the stationary states themselves and not with the transitions.[26]

Slater was a visitor at Bohr's Institute in the early part of 1924. The joint paper published by Bohr, Kramers and Slater in May 1924 was written entirely by Bohr and Kramers.[27] Slater apparently had been opposed to the idea of statistical conservation, and his intention had been that the field of virtual radiation should function as a mechanism determining the emission and trajectory of photons as real entities satisfying the classical conservation laws.[28]

In the Bohr–Kramers–Slater paper it is argued that 'spontaneous' transitions between stationary states of an atom are induced by the virtual radiation field produced by the atom, and that 'induced' transitions are occasioned by the virtual radiation produced by other atoms. The connection between transitions in one atom and those in other atoms is purely statistical, and consequently there is no conservation of energy and momentum in individual emission and absorption processes. If the frequency of the virtual radiation incident upon an atom is close to the frequency of one of the virtual oscillators corresponding to one of the possible transitions, then the atom will emit secondary virtual radiation of large amplitude, the intensity of the virtual radiation field being increased or diminished depending on whether the incident waves and secondary waves are in phase or not. The probability of an induced transition will be determined by the intensity of the virtual radiation whose frequency corresponds to the transition in question.[29]

The authors argued that Compton scattering of radiation by free electrons is a *continuous* process to which each of the illuminated electrons contributes through the emission of coherent secondary waves, each illuminated electron possessing a certain probability of taking up in a certain time a discrete amount of momentum in a certain direction – a *discontinuous* process. In order to account for the increased wavelength of the scattered radiation, they were forced to assume that the velocity of the virtual oscillators is very much greater than that of the scattering electrons – an assumption that Bohr did not regard as unduly disturbing, since the hypothesis of virtual oscillators was purely formal. He writes:

> In view of the fundamental departures from the classical space-time description, involved in the very idea of virtual oscillators, it seems at the present state of science hardly justifiable to reject a formal interpretation as that under consideration as inadequate. On the contrary, such an interpretation seems unavoidable in order to account for the effects observed, the description of which involves the wave-concept of radiation in an essential way.[30]

The wave theory is essential for the explanation of the Compton effect, Bohr believed, because the fundamental phenomenon is an observable increase in the wavelength of the scattered radiation, an increase the measurement of which presupposes the wave theory. Moreover, Bohr regarded Compton's interpretation of his experimental findings in terms of the photon hypothesis as (*pace* Compton) no less formal than his own. It is no wonder that at this period Bohr thought that the quantum theory did not admit of a realistic interpretation.

It would be wrong to regard Bohr's rejection of the strict conservation laws as simply an *ad hoc* artifice for saving the wave model of radiation, if by '*ad hoc* hypothesis' is meant one which (*a*) lacks independent theoretical and empirical support – i.e. the sole reason for introducing it is to save some theory from refutation by some data, and it is not testable independently of that data; and (*b*) it does not cohere well with the basic ideas and assumptions of the theory in question. Judged by these criteria, Bohr's hypothesis was not *ad hoc*. Firstly, it helped explain many difficulties, not only in the theory of radiation, such as the Stark effect, but also in the theory of matter, such as the Ramsauer effect.[31] For example, Bohr suggested that there are two sorts of processes of interaction, viz. those processes for which an inverse process exists ('reciprocal processes') and those for which no inverse process exists ('irreciprocal processes'). He argued that for irreciprocal interactions the strict conservation laws do not hold. In collisions between atoms and high-velocity particles, in which the duration of the collision is very short in relation to the period of motion of the bound electrons, strict conservation breaks down. The hypothesis of statistical conservation, moreover, was independently testable, as the Bothe–Geiger experiment shows. Secondly, the hypothesis cohered well with a view which Bohr had held for many years, viz. that the energy of an atom is not well defined for durations shorter than the periods of motion of the bound electrons.[32] It also chimed in with his long-standing view that the quantum of action necessitated fundamental revisions in classical mechanics; and on that assumption it was reasonable to question the strict validity of the conservation laws.

Einstein related his objections to the theory in a letter to Ehrenfest, noting that the idea of statistical conservation was 'an old acquaintance' of his. To paraphrase his objections: (*a*) Nature seems to adhere strictly to conservation laws. Why should action at a distance be an exception? (*b*) A box with reflecting walls containing radiation (in empty space free of radiation) would have to carry out an ever-increasing Brownian motion. (*c*) A final abandonment of strict causality is very hard to accept. (*d*) One would almost have to require the existence of a virtual acoustic (elastic) radiation field for solids; for it is not easy to believe that quantum theory necessarily requires an electrical theory of matter as its foundation. (*e*) The occurrence of ordinary scattering of light (not at the proper frequency of the molecules) fits badly into the scheme.[33]

Wolfgang Pauli also wrote Bohr, informing him of Einstein's objections. Einstein wished to attribute 'a greater reality' to the photon as the carrier of energy and momentum than to the wave field. Once the idea of a correspondence between the frequency of light and the frequency of the emitting electron is abandoned, the undulatory character of light seemed to him a 'secondary and indirect' effect. Pauli had objections of his own against Bohr's theory: the main one was that it entailed the existence of two distinct sorts of fluorescent radiation; and he was convinced that there was no such distinction. Besides, he held that light-quanta were no less – and no more – physically real than electrons: just as it would be unjustifiable to doubt the existence of electrons because their trajectories within atoms cannot be defined, so it would be wrong to doubt the existence of light-quanta because their trajectories in interference phenomena cannot be defined.[34] Pauli, a godson of Ernst Mach, was a positivist in his philosophy of science, and regarded the photon hypothesis as a heuristic fiction; thus he was not tempted, like Bohr, to endow photons (or electrons for that matter) with physical reality. A positivist who holds an instrumentalist view of models need not be troubled much by wave-particle duality: he may employ incompatible fictions without compunction, provided they are heuristically useful.

While some physicists were hostile to Bohr's theory, others, including Max Born and Erwin Schrödinger, were enthusiastic about it; it certainly evoked enormous interest. The theory, moreover, was testable, since it entails that in Compton scattering there is a certain probability that an electron will take up in unit time a finite amount of momentum in any given direction.[35] According to the photon theory, however, the direction of the recoiling electron is strictly, and not merely statistically, determined. This suggested to Bothe and Geiger, and to Compton and A.W. Simon, a 'crucial' experiment. Bothe and Geiger estimated that, given the conditions of their experiment, there should be no coincidences between recoil electrons and scattered photons according to Bohr's interpretation, whereas according to the photon model there should be about one recoil electron for every ten scattered photons. In their experiment Bothe and Geiger used two Geiger-counters, one to count recoil electrons and the other to count scattered photons. They discovered that for every recoil electron counted there were eleven scattered photons, a result which was highly improbable if due to chance coincidences.[36]

The possibility of testing Bohr's hypothesis of statistical conservation of energy, Compton recalls, had been suggested by W.F.G. Swann in conversation with Bohr in November 1923, before Bohr had heard of Slater's idea.[37] In their experiment Compton and Simon employed a cloud-chamber into which they passed a beam of X-rays. Out of about thirteen hundred stereographic photographs they found that, among the last 850, 38 showed both recoil electrons and scattered X-rays. In 18 cases the observed angle of scatter was within 20 degrees of value calculated according to the Compton interpretation. This incidence, they held, was four times as great as could be expected according to the Bohr–Kramers–Slater theory. While about half the random scatterings were predictable, the remaining half were unexpected, though they could plausibly be explained as being due to stray X-rays and beta rays of radio-active origin. Compton and Simon were able to conclude, then, only that a large proportion of the scattered X-rays proceed in directed quanta of radiant energy – a result which nevertheless favoured the photon theory.[38]

These experiments cast an illuminating light on Lakatos's thesis that there are no crucial experiments in science: 'crucial' is an honorific epithet that is conferred on experiments generally long after they are performed, and only after one of the rival theories has come to be regarded as empirically regressive and the other empirically progressive. Many physicists, however, regarded the experiments of Geiger and Bothe, and Compton and Simon, as crucial both before and immediately after the results were known; Bohr, interestingly enough, was among them. Some, however, did not regard these experiments as crucial. Born, for example, attempted to explain the results in such a way that the Bohr–Kramers–Slater theory was saved from refutation. It is obvious from the experiments that the results do not decisively refute that theory, though they strongly disconfirm it. As Duhem stressed, one can usually save a theory from refutation by putting the blame, as it were, on hypotheses other than those one primarily wishes to test, or by introducing into the theory additional hypotheses which transform the counter-evidence into positive evidence.

On hearing of the Bothe–Geiger results Bohr wrote R.H. Fowler: 'It seems therefore, that there is nothing else to do than to give our revolutionary efforts as honourable a funeral as possible.'[39]

2.5
The failure of spatio-temporal pictures

The Bothe–Geiger and Compton–Simon experiments were widely regarded as providing further confirmation of the photon hypothesis. Nevertheless, although these experiments inclined Bohr to abandon the Bohr–Kramers–Slater theory, they did not induce him to accept the photon hypothesis.

Geiger personally informed Bohr of his results a day before they were made public. Bohr's letter to Geiger is very revealing:

> I was quite prepared that the view we proposed on the independence of the quantum process in widely-separated atoms should turn out to be incorrect. The whole thing was more the expression of an endeavour to obtain as great as possible an application of the classical concepts than a completed theory. Not only were Einstein's objections very disturbing, but I have also felt recently that an explanation of collision phenomena, particularly Ramsauer's results on the penetration of slow electrons through atoms, presents difficulties for our ordinary space-time description of nature similar to those presented by a simultaneous understanding of interference phenomena and a coupling through radiation of the changes of state of widely-separated atoms. In general I believe that these difficulties so thoroughly rule out the retention of the ordinary space-time description of phenomena, that in spite of the existence of coupling, conclusions concerning a possible corpuscular nature of radiation lack a sufficient basis.[40]

What the Bothe–Geiger results show for Bohr is not that light is corpuscular, but rather that a classical spatio-temporal description of the interaction between radiation and matter is impossible. In a letter of 1 May 1925 to Born he writes:

> ... I should like to stress that I am of the opinion that the assumption of a coupling through radiation between changes of state in distant atoms precludes the possibility of a simple description of the physical occurrences by means of visualisable pictures ... I had come to suspect that even for collision phenomena such pictures are of even more limited applicability than is ordinarily supposed. This is of course almost a purely negative assertion, but I

feel that, particularly if the coupling should really be a fact, one must have recourse to symbolic analogies to an even greater extent than hitherto. Just recently I have been racking my brains to dream up such analogies.[41]

In a letter of 28 January 1926 to Slater, Bohr writes:

Although of course we were wrong in Copenhagen as regards the coupling of the quantum processes – in which respect I have a bad conscience in persuading you to our view – I believe that Einstein agrees with us in the general ideas, and that especially he has given up any hope of proving the correctness of the light quantum theory by establishing contradictions with the wave theory description of optical phenomena. In my opinion the possibility of obtaining a space-time picture based on our usual conceptions becomes ever more hopeless.[42]

It is doubtful whether Bohr is right about Einstein, who was still devising experiments to disconfirm the existence of spherical radiation waves.[43] Bohr, it is clear, was still not prepared to grant the 'correctness' of the photon hypothesis. Slater's reply of 27 May 1926 is interesting. He writes:

As far as radiation is concerned, you need not have a bad conscience for having persuaded me to think that there were no quanta. I think we did a useful service by coming out with the definite suggestion that there were none; for that called peoples' attention to the fact that there were then no experiments which could not be perfectly well explained without assuming them.[44]

Whereas Slater was inclined to regard the Bothe–Geiger and Compton–Simon results as further evidence of the existence of photons, Bohr regarded these results as showing, not that the photon theory is correct and the wave theory wrong, but rather that there can be no coherent classical model of the interaction between radiation and matter.

Bohr acknowledged the refutation of the Bohr–Kramers–Slater theory by the Bothe–Geiger experiment in a postscript, dated July 1925, to his paper on collision phenomena. In this postscript he repeatedly stresses the formal character of mechanical models and the difficulties in making use of visualisable pictures:

It must at the same time be emphasised that the question of a coupling or of an independence between the individual, observable atomic processes cannot simply be regarded as a distinction between two well-defined conceptions of the propagation of light in empty space, which correspond perhaps to a corpuscular theory or to a wave theory of light. It concerns, rather, the problem to what extent the space-time pictures, by means of which natural phenomena have hitherto been described, are applicable to atomic processes.[45]

Bohr's point is that if radiation *in vacuo* consists of waves, then no spatio-temporal description of the process of interaction between these waves and matter is possible in terms of a visualisable model. We can expect, Bohr thinks, that the generalisation of the classical electrodynamic theory will require 'a thoroughgoing revolution in the concepts upon which the description of nature has rested up to now.' We must be prepared to discover further processes 'which are in strange contrast to ordinary space-time pictures, as strange as the contrast between the coupling of individual processes in distant atoms and the wave description of optical phenomena'.[46] Part of that thoroughgoing conceptual revolution will be the overthrow of our ordinary spatio-temporal models or at least a recognition that there are strict limits to their meaningful applicability. Whether this will make room for new models Bohr does not say.

2.6
Discontinuity and unvisualisability

In November 1925 Bohr elaborated an address entitled, 'Atomic Theory and Mechanics', which he had delivered on 30 August to the Sixth Scandinavian Mathematical Congress at Copenhagen. He observed there that the conception of 'light-quanta of energy $E = h\nu$' is of a formal nature, since the definition and measurement of the frequency 'rests exclusively on the ideas of the wave theory.'[47] Apropos the Bothe–Geiger and Compton–Simon experiments he writes:

> From these results it seems to follow that, in the general problem of the quantum theory, one is faced not with a modification of the mechanical and electrodynamical theories describable in terms of the usual physical concepts, but with an essential failure of the pictures in space and time on which the description of natural phenomena has hitherto been based.[48]

What Bohr means is that the difficulties in the interpretation of the quantum theory are not due to inadequacies in the mechanical concepts of the theory, but to inadequacies in the models used for its interpretation. This statement is virtually identical with one made in 1922.[49] What these experiments impressed on Bohr was not so much a recognition of wave–particle duality as a recognition of the failure of the classical models. Indeed, he plays down the duality and constantly stresses the 'insufficiency of mechanical pictures' and the 'limitation on our usual means of visualization'.[50]

Bohr regarded the failure of the classical models of quantum processes as being due, not to the microscopic dimensions of the phenomena, but rather to the breakdown of continuity presupposed by the quantum theory, to 'an element of discontinuity in the description of atomic processes quite foreign to the classical theories'.[51] He writes:

> The more precise formulation of the content of the quantum theory appears, however, to be extremely difficult when it is remembered that all the concepts of previous theories rest on pictures which demand the possibility of a continuous variation.[52]

Continuity is presupposed both by the classical electromagnetic theory, according to which radiation is a continuous distribution of energy in space, interacting in a continuous way with matter, and by classical electrodynamics, mechanics and thermodynamics, according to which energy, although not continuously distributed in space, is exchanged in a continuous manner between material systems. Since the quantum theory presupposes discontinuity, it may not be possible to provide a comprehensive and consistent interpretation of it in terms of the models appropriate to the classical theories.

It is not exactly clear what Bohr means when he speaks of 'an essential failure of the pictures in space and time' and 'the renunciation of mechanical models in space and time'.[53] I believe that he has in mind a partial, rather than a complete, failure, for he talks also of the 'insufficiency of mechanical pictures' and 'a limitation of our usual means of visualization'.[54] In what respect are the classical models limited? Largely in so far as they do not provide a completely comprehensive conception of the *behaviour* of the entities they represent, i.e. of all their actions and reactions. The models imply, contrary to fact, that the objects have continuous trajectories in space and time. The contexts in which the classical models are said to fail mainly concern interaction processes, such as interactions between matter and radiation, collisions between atoms and electrons etc. If this is Bohr's view, then the wave model might be said to be applicable to radiation propagating in free space, but not to radiation in interaction with matter. Moreover, whereas the particle model is formally applicable in the latter case, it does not provide a realistic representation of what is taking place.

This interpretation of Bohr's statements on the failure of the classical models coheres with his deep-seated disinclination to reject the wave model of radiation and with his inclination to locate the source of the difficulty in mechanics. It is clear that in 1925 his response to the refutation of the Bohr–Kramers–Slater theory was not a naive acceptance of wave-particle duality – the existence of two incompatible, but equally well-confirmed, theories of radiation – but rather an even more acute recognition of the limited applicability of the classical models of radiation. His response was to retreat even further from a realist construal of the quantum theory: the theory, for the time being, must be limited to 'saving the phenomena'.

3
From duality
to complementarity

Bohr's response in 1925 to the failure of the Bohr–Kramers–Slater theory was, as we have seen, to regard wave–particle duality as having a formal, rather than realistic, significance; he saw it as a 'limitation of our usual means of visualization'.[1] That same year, however, wave–particle duality was generalised by Louis-Victor de Broglie to cover the theory of matter. Moreover, the advent of quantum mechanics (i.e. matrix mechanics and wave mechanics), rather than dispelling the duality problem, heightened it. How did Bohr respond to these new developments in the quantum theory?

3.1
A matter of waves

In the autumn of 1923 de Broglie put forward the extremely bold hypothesis that any particle whatever is associated with a wave whose wavelength λ and frequency ν are proportional to the particle's momentum and energy respectively: $\lambda = h/p$, $\nu = E/h$. De Broglie regarded his hypothesis as the basis of a new mechanics, a wave mechanics related to classical mechanics in the same way as wave optics is related to ray optics.[2] On this view the trajectory of a particle corresponds to the ray of the wave associated with it (i.e. the normal to the equiphase surface of the wave), and the velocity of the particle corresponds to the group velocity of the wave. De Broglie suggested, moreover, that in certain circumstances particles should exhibit diffraction effects: a stream of electrons, for example, which passes through a slit in a screen should undergo diffraction.[3]

When Einstein got to know about de Broglie's work in 1924, he saw at once that it had a bearing on his current work on the quantum statistics of an ideal gas. In June 1924 Einstein was sent a paper by the Indian physicist, S.N. Bose, who had derived Planck's radiation law by treating radiation as if it were an ideal gas consisting of photons subject to a statistical theory different from that of Boltzmann. Whereas Boltzmann's theory related to mutually independent, distinguishable particles, Bose's theory had to do with non-independent, indistinguishable particles.[4] Bose's statistical theory was essentially the same as the one that Planck had used in his original derivation of his radiation law.

Einstein had already argued, as far back as 1907, that the quantum hypothesis had implications for the molecular theory of heat – for example that the specific heats of liquids and solids should approach zero as their temperature drops towards zero.[5] Einstein now hit upon the idea that the Bose statistics might be the correct statistics to use in the molecular theory of heat. Elaborating on this idea, he showed that these statistics give a correct account of the specific heat of an ideal gas in the high density/low temperature region.[6] There is, he noted, 'a far-reaching formal relationship between radiation and gas': in each case the classical theories of heat (e.g. the Rayleigh–Jeans law, the Boltzmann statistics) fail in the high density/low temperature region.[7] Einstein had been aware of this analogy since 1905: he had argued then that high-frequency radiation behaves in some respects like an ideal gas. Stimulated by de Broglie's work, he now turned the hypothesis the other way round, arguing that an ideal gas in some respects behaves like radiation. He showed that the mean–square energy fluctuation $\langle \Delta \epsilon_s^2 \rangle$ of the ideal gas calculated according to the new statistics is the sum of two terms, n_s and n_s^2/z_s, the former alone being present for a gas consisting of mutually independent particles. The second term could be interpreted as being due to the fluctuating superposition of waves. There is an exact analogy here with the formula for the mean–square energy fluctuations of black-body radiation that Einstein had discovered in 1909. There is, he believed, 'more than a mere analogy involved here'.[8] Thus just as radiation possesses some of the characteristics of particles, so particles may be expected to possess some of the characteristics of radiation.

Having Einstein's blessing, de Broglie's work was not ignored. Walter Elsasser argued that the experiments of C.J. Davisson and C.H. Kunsman, which showed anomalies in the scattering of electrons from metal plates, might be interpreted as diffraction effects, thus confirming de Broglie's prediction.[9] More striking evidence of electron diffraction was forthcoming in 1927, in the experiments of Davisson and L.H. Germer, and of G.P. Thomson and Alexander Reid. Davisson and Germer showed that a beam of electrons is reflected from a nickel crystal at angles predicted by the Bragg formula for X-ray diffraction, $\lambda = (2d \cos \theta)/n$ ('d' signifying the distance between the crystal planes, 'θ' the angle of incidence). The value of the de Broglie wavelength of the electrons, calculated from the Bragg formula (taking refraction into account), is in close agreement with the value obtained by way of the de Broglie formula $\lambda = h/p$.[10]

How did Bohr view the hypothesis of matter waves? Unlike de Broglie and perhaps Einstein, he regarded it as purely formal, of no realistic significance. In the postscript to his paper of 1925 on collision processes he remarks:

> The renunciation of space-time pictures is characteristic of the formal treatment of the problems of the theory of radiation and the mechanical theory of heat which has been attempted in recent publications of de Broglie and Einstein.[11]

During the early part of 1926, however, the hypothesis of electron spin seems to some extent to have revived Bohr's confidence in the usefulness of mechanical models in the quantum theory.

In 1923 Bohr had argued that the puzzling multiple splitting of spectral lines in a magnetic field (the anomalous Zeeman effect) could not be explained in terms of mechanical models. He suggested that the effect was due to '. . . a mechanically undescribable "strain" in the interaction of the electrons which prevents a unique assignment of quantum indices on the basis of mechanical pictures'.[12] In 1925 Pauli proposed an explanation in terms of a fourth quantum number representing a 'peculiar, classically undescribable sort of two-valuedness' in the quantum-theoretical properties of the electron.[13] He eschewed interpreting this notion in classical mechanical terms.

Later that year Samuel Goudsmit and G.E. Uhlenbeck suggested that Pauli's fourth quantum number represented an intrinsic angular momentum of the electron.[14] This hypothesis was not free of difficulties: Goudsmit's and Uhlenbeck's calculations of the doublet separation were too large by a factor of two; and the rotational velocity of the surface of the spinning electron was much in excess of the velocity of light. Heisenberg and Pauli rejected the hypothesis, Pauli regarding it as a retrogressive step, a throwback to the defunct classical models. This response was to some extent an expression of his sympathy with a positivistic theory of science.[15]

Bohr too was initially critical of the spin hypothesis. He changed his mind, however, after discussing it with Einstein at Leyden in December 1925. He asked Einstein why there should be a coupling between the spin axis and the orbital motion of the electron. Einstein's reply, that the coupling is an immediate consequence of the special theory of relativity, came as a complete revelation to him, convincing him of the usefulness of the spin hypothesis. In a letter to *Nature* in support of the hypothesis he writes:

> In my article expression was given to the view that these difficulties were inherently connected with the limited possibility of representing the stationary states of the atom by a mechanical model. The situation seems, however, to be somewhat altered by the introduction of the hypothesis of the spinning electron which, in spite of the incompleteness of the conclusions that can be derived from models, promises to be a very welcome supplement to our ideas of atomic structure . . . Indeed it opens up a very hopeful prospect of our being able to account more extensively for the properties of elements by means of mechanical models, at least in the qualitative way characteristic of applications of the correspondence principle.[16]

3.2
Quantum mechanics and the correspondence principle

Between 1913 and 1922 the quantum theory of atomic structure had made remarkable progress: it explained an impressive amount of spectroscopic data. The theory, however, lacked coherence: it was something of a hotch-potch of diverse rules of thumb, problems being solved by a piecemeal application of classical mechanics and electrodynamics suitably modified by quantum conditions. Moreover, after 1922 the theory faced serious problems, such as that of determining the energy states of the helium atom, the anomalous Zeeman effect, etc. Attempts to solve these problems by constructing semi-classical, mechanical models of the atom failed. By 1925 there was a general feeling that the quantum theory was in serious difficulties. No one was more acutely aware of the difficulties than Bohr and Pauli. In a letter to Ralph Kronig, Pauli complains about the muddled state of physics and expresses the hope that Bohr will come to the rescue with a new idea.[17]

It was not Bohr, however, who developed the new idea which resuscitated the quantum theory, but his young assistant, Heisenberg. In the summer of 1925 Heisenberg discovered the basis of a more coherent and comprehensive quantum theory, matrix mechanics, in which the position co-ordinates and momenta of bound electrons are treated mathematically as matrices, the elements of which correspond to transition amplitudes. Although the new idea was not provided by Bohr, Heisenberg had arrived at his idea as a consequence of his work on the dispersion theory of Kramers, a theory whose origins lay in the Bohr-Kramers–Slater theory of 1924.[18] Heisenberg's ideas were rapidly developed by Born and Pascual Jordan, who formulated them explicitly as matrices. The following year, 1926, Schrödinger, elaborating upon de Broglie's theory of matter waves, invented wave mechanics.[19]

What was Bohr's attitude to these new theories? In 1922 he had been hopeful that a future theory would be found which, while retaining the characteristic features of the quantum theory, could be regarded as a rational generalisation of classical electrodynamics.[20] In 1926 he regarded quantum mechanics as the fulfilment of that hope. He saw matrix mechanics as 'a precise formulation of the tendencies embodied in the correspondence principle'.[21] What exactly did Bohr mean by this? In order to answer this question we shall have to take a closer look at the correspondence principle and Bohr's interpretation of it.

It is a consequence of Bohr's theory of the hydrogen atom that in the limiting region of high quantum numbers the difference between the energies of consecutive stationary states converges to zero. Moreover, as a consequence of Bohr's postulate, $h\nu = E_n - E_m$, relating the energy $h\nu$ of the emitted radiation to the difference $E_n - E_m$ between the energies of the stationary states involved in the transition, the frequency ν of the emitted radiation tends towards zero as the energy difference converges. In this limiting region the frequency of the emitted radiation coincides asymptotically with the frequency of revolution of the electron, or with the frequency of a harmonic component of the revolution frequency in the case of transitions between electron orbits which are not consecutive. Moreover, the intensity of the emitted radiation corresponds asymptotically to the amplitude of the harmonic component of the periodic motion of the electron; and this amplitude gives a measure of the probability of the occurrence in question.[22] Thus, in this limiting region not only do the results of classical electrodynamics and Bohr's quantum theory of the emission and absorption of radiation asymptotically coincide, but also the classical model of radiation emission and absorption becomes applicable. This consequence of Bohr's theory agrees with the earlier finding that the classical theory and model of radiation is adequate for radiation of low frequency.

Bohr originally called this asymptotic relation between classical electrodynamics and the quantum theory an 'analogy'. In a letter to Rutherford of 21 March 1913 he speaks of 'the most beautiful analogi [*sic*] between the old electrodynamics and the considerations used in my paper'.[23] He calls this analogy a 'correspondence' and talks of the 'correspondence principle' in his paper of 1920, 'Über die Serienspektren der Elemente':

> Moreover, although the process of radiation can not be described on the basis of the ordinary theory of electrodynamics . . . there is found, nevertheless, to exist a far-reaching *correspondence* between the various types of possible transitions between the stationary states on the one hand and the various harmonic components of the motion on the other hand. This correspondence is of such a nature, that the present theory of spectra is in a certain sense to be regarded as a rational generalisation of the ordinary theory of radiation.[24]

The correspondence principle is a generalisation of these implications of Bohr's theory of the hydrogen atom: it states that in general, for high quantum numbers, the results yielded by the quantum theory must coincide approximately with those derived from the classical theories. Bohr intended the general principle to operate as a methodological rule to guide the formulation of laws in the quantum theory. It is a tribute to his genius that the principle was used with brilliant success for a period of ten years.

Bohr's justification for the principle was based upon his conception of the relation of the quantum theory to the classical theories. In his Nobel Prize Address of 1922 he states that, in spite of its fundamental departure from the postulates of classical mechanics and electrodynamics, the quantum theory is a natural generalisation of classical electrodynamics.[25] What makes the one theory a generalisation of the other is the relation of correspondence between them. Correspondence in Bohr's sense is a species of reduction relation. A theory T_2 may be said to correspond in Bohr's sense, partially or completely, to another theory T_1 if and only if some or all of the laws (i.e. law-like generalisations) of T_2 reduce to the laws of T_1 when certain parameters of T_2 are given limiting or extreme values. What it means for the laws of T_2 to reduce to the laws of T_1 is not that the laws of T_2 should be transformed into, or become identical with, the laws of T_1, but rather that the laws of T_2 should yield the same numerical values as the laws of T_1 under a certain assignment of values to the constants of T_2. Under the reduction relation the constants of T_2 do not disappear; nor do its laws alter in form. The laws of T_2 and T_1, *qua* mathematical formulas, remain distinct, although they yield identical numerical solutions. If correspondence in this sense obtains, then T_1 may be said to be reducible to T_2, and T_2 may be said to correspond to T_1. Quantum mechanics reduces to classical mechanics when Planck's constant h is given the value zero, just as the special theory of relativity reduces to classical mechanics when the light-velocity constant c is given the value ∞. Quantum mechanics was for Bohr the generalisation of classical mechanics that he had long been seeking.

Quantum mechanics, however, does not reduce to classical mechanics in the sense that the latter theory is logically derivable from the former *per se*; what is logically derivable from quantum mechanics *per se* is an approximation to classical mechanics in a limiting region. Classical mechanics is logically derivable only on the counter-factual assumption that the quantum of action h has the value zero – an assumption which flatly contradicts the fundamental assumption of quantum mechanics that the quantum of action has a finite, non-zero value. This fact alone, of course, does not entail that classical mechanics and quantum mechanics are logically incompatible theories in the sense that they are logical contraries, i.e. they cannot both be true, though they may both be false. But if classical mechanics entails or logically presupposes that the value of h is zero, and quantum mechanics entails or presupposes the contrary assumption, that h has a definite, non-zero value, then the two theories are logically incompatible. Classical mechanics, however, does tacitly presuppose that $h = 0$, and quantum mechanics explicitly presupposes that $h = n \neq 0$; hence the two theories are logically incompatible. The principal constants of a theory, and the values assigned them by the theory, are, if anything is, essential characteristics of the theory. It is, then, strictly incorrect to say that quantum mechanics 'logically contains' classical mechanics.

Bohr, I believe, recognised the incompatibility between classical mechanics and quantum mechanics. He frequently stressed that the fact that the quantum theory is a generalisation of electrodynamics does not mean simply that electrodynamics is a limiting case of a vanishing quantum of action. Thus he writes:

> . . . in the limiting region of large quantum numbers there is in no wise a question of a gradual diminution of the difference between the description by the quantum theory of the phenomena of radiation and the ideas of classical electrodynamics, but only of an asymptotic agreement of the statistical results.[26]

Bohr stressed that the quantum postulate does not lose its significance in the limit of high quantum numbers, and that the conceptions of stationary states and discontinuous transition processes retain their validity in the limiting region.[27] The mechanisms of radiation emission and absorption are radically different according to the two theories, even in the limiting region in which the two theories are formally (approximately) equivalent. In the classical case emission is a continuous process in which the atom emits radiation of the fundamental frequency and of the higher harmonic frequencies of the periodic motion of the electron; in the quantum case the emission process is discontinuous, and the frequency of the emitted radiation, which is monochromatic, is not causally determined by the periodic motion of the electron.

Bohr's notion of correspondence is of great importance for the concept of progress in physics; its significance is not confined to the relation between the quantum theory and the classical theories. Indeed, a radically new theory in physics generally stands in a relation of Bohrian correspondence to the theory which it supersedes. For example ray optics is reducible to wave optics on the assumption that wavelengths are infinitely short; and the special theory of relativity reduces to classical mechanics on the assumption that v/c equals zero. In general a new theory T_2 supersedes an old theory T_1 only if T_2 explains all the phenomena explainable by T_1 as well as phenomena unexplainable by T_1.[28] This condition is usually satisfied through correspondence: the laws of the new theory T_2 are generalisations of the laws of the old theory T_1, in the sense in which the van der Waals law,

$$(P + a/V^2)\ (V - b) = \text{const.} \times T,$$

is a generalisation of Boyle's law, $PV = \text{const.} \times T$; the former formula reduces to the latter on the assumption that there are no intermolecular forces and that the molecules are mass-points (i.e. $a = 0$, $b = 0$).

The correspondence relation, moreover, provides the grounds for an explanation not only of the failure of the superseded theory but also of its limited success. The superseded theory can generally be shown to fail because it contains, often tacitly, an ideal assumption (idealisation) which is false; it ignores some important factor which the new theory brings to light and corrects, often by introducing a new parameter. T_2 reveals the degree of idealisation, as it were, of the previous theory T_1.[29] The limited success of the superseded theory can usually be shown to be due to the fact that the factor which it neglects is empirically negligible under certain conditions. The new theory corrects the old one by exposing the falsity of one of its tacit idealisations; it is also more general in so far as it has (it is presumed) fewer ideal, counter-factual assumptions than the preceding theory, and also in so far as its results are approximately correct for a wider range of phenomena than are those of the older theory.

Classical mechanics is an idealisation in the sense (among others) that it presupposes that the quantum of action is zero. Similarly, our 'forms of perception', and many of the classical physical concepts, are idealisations in that they are valid only under special conditions. After 1927 Bohr emphasised what he saw as a profound similarity between the theory of relativity and the quantum theory: in both theories we are concerned with laws which 'lie outside the domain of our ordinary experience and which present difficulties to our accustomed forms of perception.'[30] Just as the classical, sharp distinction between space and time rests upon the smallness of the velocities ordinarily met with compared with the velocity of light, so the appropriateness of the (classical) causal, space-time description depends upon the small value of the quantum of action compared with the actions involved in ordinary sense perception.[31] The classical theories are idealisations in the sense that they hold under the ideal assumptions of the infinitely large value of the velocity of light and the infinitely small value of the quantum of action. Bohr's point here is one that was stressed by his friend, the philosopher Harald Høffding:

> . . . continuity, causality, time and space – as conceived by Kant – possess an ideal perfection to which there is no corresponding experience. Continuity is an idea to which experience only gives us approximations. What Kant calls forms are, as a matter of fact, abstractions and ideals which, in accordance with the nature of our knowledge, we set up and use as measures and rules for our inquiries.[32]

Between 1913 and 1925 Bohr's approach to the development of the quantum theory was firmly based on the correspondence principle. He saw that for processes involving high energies and taking place on a comparatively large scale – i.e. for macroscopic processes – classical mechanics is approximately correct in the sense that for the most part the errors to which it gives rise are not readily detectable. He was convinced that the quantum theory should be developed by trying to discover suitable restrictions on the laws of classical mechanics that took account of the quantum of action. What was not required was a wholesale replacement of the concepts of the classical theories. In the early twenties, however, he realised that introducing various restricting conditions into classical mechanics was too piecemeal an approach, and that what was required was a new set of laws governing the use of the classical concepts, a unified and systematic generalisation of classical mechanics. Quantum mechanics he saw as the fulfilment of that wish.

Although matrix mechanics and wave mechanics are formally equivalent, as Schrödinger showed in the spring of 1926, the two theories are very different: matrix mechanics employs discrete quantities and a non-commutative algebra; wave mechanics employs continuous quantities and differential equations. Bohr noted that matrix mechanics does not provide 'a space-time description of the motion of atomic particles'; unlike wave mechanics, matrix mechanics does not lend itself so readily to interpretation in terms of a visualisable model; Bohr hoped that it would provide a way of avoiding the difficulties associated with the use of such models. Although he welcomed Schrödinger's theory, he did not accept Schrödinger's realistic interpretation of it in terms of the wave model.

Schrödinger took the analogy between wave optics and wave mechanics to be more than formal: he conceived of matter and radiation as essentially undulatory entities, which appear to be corpuscular when their wavelength is small in relation to the dimensions of the path of their propagation, particles being wave-packets, concentrations formed by the superposition of individual waves. He interpreted the amplitude function ψ (corresponding to the optical amplitude function ρ in wave optics) in electrodynamical terms: $\psi\psi^*e$ (with ψ^* the complex conjugate) is a measure of the density of the electric charge e. He argued that there are in fact no discrete energy levels and no discontinuous transitions between stationary states, apparent discontinuities being simply changes from one vibrational mode to another, analogous to 'beats' in acoustics.[33]

Schrödinger visited Bohr's institute at the beginning of October 1926. He, Bohr, Heisenberg and others discussed the interpretation problem intensively. Against Schrödinger, Bohr argued that there can be no elision of discontinuity. He held, however, that there is a place, indeed an essential place, for continuity in quantum mechanics. The discussions convinced him that the continuity–discontinuity duality was reflected in wave mechanics and matrix mechanics respectively, a view which agreed with his long-held opinion that discontinuity must be assumed to hold for the interaction between radiation and matter, and continuity for the propagation of radiation *in vacuo*. This is not to say that he regarded wave mechanics and matrix mechanics as simply reflecting wave–particle duality, i.e. that wave mechanics is required for the explanation of undulatory phenomena and matrix mechanics for corpuscular phenomena; for he held no such view.

Bohr unequivocally rejected Schrödinger's interpretation of wave mechanics. Indeed, he immediately accepted Born's probabilistic interpretation, which was published just before Schrödinger's visit to Copenhagen.[34]

In fact, wave mechanics, just as the matrix theory, on this view represents a symbolic transcription of the problem of motion of classical mechanics adapted to the requirements of quantum theory and only to be interpreted by an explicit use of the quantum postulate.[35]

By 'symbolic' here Bohr means 'purely formal'. In an earlier typescript version of this passage, after the words 'the quantum postulate', we find the phrase 'and the concept of material particles'.[36] Elsewhere in the published article Bohr was more explicit:

> In judging the possibilities of observation it must, on the whole, be kept in mind that the wave–mechanical solutions can be visualized only in so far as they can be described with the aid of the concept of free particles.[37]

3.3
The continuity-discontinuity duality

It is clear that in the autumn of 1926 Bohr still regarded the classical models of wave and particle as formal or 'symbolic' and not necessarily of any realistic significance. On his return to Zurich after his famous visit to Bohr in Copenhagen, Schrödinger wrote Bohr:

> For you seem to me to find a kind of provisional resting place in the view that the complete, apparently visualisable pictures are in reality only to be taken as symbolic . . . But I am not able wholly to soothe myself with this provisional solution. It seems to me as little *generally* applicable as my own.[38]

Unfortunately Bohr's lecture of 17 December 1926 to the Danish Academy, entitled 'Atomic Theory and Wave Mechanics', was not published, and no manuscript has survived.[39] Høffding, however, described the lecture to Emile Meyerson:

> . . . he suspects that we cannot decide whether the electron is a wave motion (in which case we could avoid discontinuity) or a particle (with discontinuity between the particles). Certain equations lead us to the former conception, certain others to the latter . . . In a conversation which he had with me after the lecture Mr. Bohr told me that he is ever more convinced of the necessity of symbolisation if we wish to express the latest findings of physics.[40]

By 'symbolisation' is meant 'a formal use of mathematical symbols', i.e. the eschewal of a realistic interpretation of the mathematical formulas of the theory.

Unfortunately Bohr published next to nothing during this crucial period. Fortunately, however, his letters reveal a good deal of what he was thinking. In a letter of 24 November to Darwin he writes:

> ... it is very interesting to see how the notion of a corpuscle or of a wave presents itself as the more convenient quite according to the place in the construction where the feature of discontinuity involved in the postulates is explicitly introduced. Indeed, this is only what might be expected, since every notion, or rather every word, we use is based on the idea of continuity, and becomes ambiguous as soon as that idea fails.[41]

Bohr wrote in exactly the same vein to Schrödinger on 2 December 1926, and again in a letter of 22 January 1927 to Bibhubhusan Ray:

> As you know, Schrödinger himself has entertained the hope that his methods should lead to the re-establishment of a proper continuity theory avoiding the quantum paradoxes. This, however, would seem to be an illusion since it appears that these paradoxes are deeply rooted in nature, and at present at any rate are unavoidable in the description of experimental facts. The dualism inherent in the quantum theory has rather been still more emphasized through de Broglie's and Schrödinger's work. The old dilemma of the light waves and light quanta has been extended to the electron itself and it is very interesting to see, how the notion of a corpuscle or a wave presents itself as the more appropriate, according to the way in which the theory is formulated, i.e. at what place the discontinuous element is explicitly [*sic*] introduced. Indeed, this is just what might be anticipated since every notion, or better every word, we use is based on the idea of continuity and becomes ambiguous as soon as that idea fails.[42]

By 'ambiguous' Bohr means 'ill-defined': 'unambiguous' is the word which Bohr usually uses for the Danish *entydig*, which literally means 'signifying one thing, unequivocal'. Bohr may have intended to say here merely that continuity is a presupposition of the meaning of every concept of classical physics rather than of every concept *sans phrase*: his point is that the classical models, such as the particle model and the wave model, presuppose continuous spatio-temporal behaviour; it is for example implicit in the concept of a classical particle that it has a continuous spatio-temporal trajectory. If the presupposition of continuity fails, then the classical physical concepts lack well-defined application: the predicates 'is a particle' and 'is a wave' are no longer well defined, and hence cannot be unequivocally applied in every situation. Bohr may mean, however, something more general, that the meaning of every word presupposes continuity in the sense that continuity is a presupposition of the conceptualisation of all experience – the Kantian idea that concepts are means of establishing continuity within our perceptual experience, an idea that was continually stressed by Høffding. In 1927, for example, Bohr states that 'every word in the language refers to our ordinary perception'.[43]

It is clear from his letters that Bohr was preoccupied with the question of duality during the winter of 1926–27. The letters to Darwin, Schrödinger and Ray indicate, I believe, a change in his attitude to wave–particle duality. In these letters he says (to paraphrase) that 'it is interesting to see how the concept of particle or the concept of wave presents itself as the more appropriate depending on the place where discontinuity is introduced'. Apparently he now thought that there were appropriate uses for the non-standard models, i.e. the wave model of matter, the particle model of radiation. He notes, moreover, that the quantum paradoxes 'are deeply rooted in nature, and at present at any rate are unavoidable'. This statement is a fairly explicit recognition of wave-particle duality: duality is *rooted*, not just in present theory, but in *nature*. We see here a sympathy towards the idea of duality which Bohr had never previously expressed; he had come to think that both the wave model and the particle model have an indispensable role to play in the interpretation of experimental data.

Bohr's view now was that when continuity obtains, the standard models are applicable, i.e. matter may be conceived of as corpuscular and radiation as undulatory; when, however, discontinuity prevails, the standard models break down, since they presuppose continuity, and the non-standard models then suggest themselves. Although there is no wholly adequate single model of matter and of radiation, and no single comprehensive interpretation of quantum mechanics, there is a partially adequate interpretation for every application of the theory. Given the theoretical situation at the time, this was a cautious and reasonable approach to the interpretation problem.

3.4
The uncertainty principle

During February 1927 Heisenberg derived from the Dirac–Jordan quantum-mechanical transformation theory the relations between the uncertainties q_1, P_1 and E_1, t_1 in the values of position q and momentum p, and of energy E and time t respectively: $q_1 P_1 \sim h$, $E_1 t_1 \sim h$. He interpreted these relations as signifying that the more precisely the position of an object is known, the less precisely is its momentum known (and vice versa), and the more precisely the energy is known, the less precisely is the time known.[44] The uncertainty principle, as Heisenberg first formulated it, is an epistemic principle: it lays down limits to what we can know; it is not an ontic principle, circumscribing the physical properties which objects may instantiate. On this view quantum mechanics is an indeterministic theory simply because the data required for deterministic predictions of the sort that classical mechanics provides are unobtainable. At this time Heisenberg was very sympathetic to a positivistic theory of science, and hence believed that the question whether quantum-mechanical objects possess exact simultaneous positions and momenta is physically meaningless. From this point of view the uncertainty principle could be given an ontic interpretation (or a semantic variant of the ontic interpretation): objects possess (or can be said meaningfully to possess) only observable properties; since the exact simultaneous position and momentum of a quantum-mechanical object is unobservable, no such object possesses (or can meaningfully be said to possess) such a property. Heisenberg was not averse to such an ontic extension of the epistemic principle.

The uncertainty principle had earlier been suggested, in an inchoate form, by Pauli: in a letter to Heisenberg of 19 October 1926 Pauli writes:

> One can look at the world with the p-eye and with the q-eye, but if one wishes to open both eyes at the same time, one goes wrong.[45]

In his reply of 28 October Heisenberg writes:

> The equation $pq - qp = -i\hbar$ thus always corresponds in the wave conception to the fact that it makes no sense to speak of a monochromatic wave at a definite instant (or a very short time interval) . . . Analogously it makes no sense to talk of a particle with a definite velocity. But if one doesn't take the velocity and position so exactly, it may very well have a sense.[46]

Bohr and Heisenberg discussed the nature of this uncertainty intensively before the end of February 1927, when Bohr departed for Norway on a skiing holiday. In a letter of 19 March 1927 Heisenberg writes to Bohr in Norway:

> These last weeks I have myself been working very energetically at carrying through the programme (in connection with the Dirac–Jordan quantum mechanics) about which we spoke before your departure . . . Moreover one can see that the transition from micro- to macro-mechanics is *now* very easy to

understand; classical mechanics is altogether a part of quantum mechanics. Apropos the old question, 'statistical or causal laws?' the position is thus: one can *not* say that quantum mechanics is statistical. But when one wishes to calculate 'the future' from 'the present' one can only get statistical results, since one can never discover every detail of the present.[47]

Heisenberg's point here is that (deterministic) classical mechanics is not simply the large-number limit of (statistical) quantum mechanics; rather the statistical character of quantum mechanics is due to a fundamental limitation on measurability which results from the discontinuities involved in the process of measurement. Classical mechanics is a special case of quantum mechanics in the sense that quantum mechanics reduces to classical mechanics when the quantum discontinuities involved in measurement have negligible effects.

Heisenberg supported his physical interpretation of the uncertainty relations by means of an imaginary experiment. The position of an electron, he argued, could in theory be accurately measured by illuminating it with light of very short wavelength, and by observing the reflected light in a microscope. Owing to the limited resolving power of lenses, the wavelength of the light would have to be as short as that of gamma radiation. The illumination of an electron by gamma rays, however, would bring about a discontinuous change in the momentum of the electron, owing to the high frequency of the rays and an accompanying Compton effect. This change would render our knowledge of the momentum of the electron uncertain. In his famous paper on the uncertainty relations Heisenberg writes:

> At the moment of the position determination, when the light-quantum is diffracted by the electron, the momentum of the electron is changed discontinuously. The shorter the wavelength of the light, i.e. the more accurate the position measurement, the greater the change in the momentum. At the moment the position of the electron is ascertained, its momentum can be known only within a magnitude that corresponds to this discontinuous change . . .[48]

Heisenberg's paper was completed at the end of February 1927 while Bohr was on holiday in Norway. On Bohr's return, intensive discussions were resumed. Bohr was very critical of Heisenberg's derivation of the uncertainty relations. Heisenberg wished to base his derivation solely upon the Dirac–Jordan theory, on the grounds that the discontinuity which he saw as essential to the quantum theory could be formulated more easily in the Dirac–Jordan theory than in wave mechanics; and he wished to attribute the uncertainties to the discontinuities inherent in the measuring process. Bohr, however, wished to employ the wave theory in the derivation and to regard the uncertainties as a consequence of wave–particle duality. He objected that Heisenberg 'had not started from the dualism between particles and waves'.[49] Whereas Heisenberg regarded wave mechanics merely as a useful tool for solving mathematical problems in quantum mechanics, Bohr 'seemed inclined to place the wave–particle dualism among the basic conceptions of the theory'.[50]

The disagreement between Bohr and Heisenberg is described in a letter of 3 May 1928 written by Klein to George Birtwistle on behalf of Bohr:

> What to him was the main point of his work was the complementarity of the wave and particle idea, which did not appear clearly from Heisenberg's work, where the origin of the uncertainty relations was sought in the quantum discontinuities rather than in the difficulties of describing particles by means of wave packages so intimately connected with the theory of optical instruments used in observations.[51]

Bohr argued that what precludes the measurement of the momentum of the electron in the 'gamma-ray microscope' experiment is not the discontinuity of the momentum change as such but rather the impossibility of *measuring* the change. What prevents measurement of the momentum change is the indispensability of the wave model for the interpretation of this experiment. The Compton–Simon experiment shows that the discontinuous change in momentum can be accurately determined, provided the angle of scatter of the incident photon can be precisely determined. In the 'gamma-ray microscope' experiment, however, the angle of scatter cannot be determined with an uncertainty which is less than the angle 2θ subtended by the diameter of the lens: it is thus impossible to tell at what angle within the angular aperture of the lens the photon is scattered; the photon may be scattered at any angle between EA and EB (see Figure 1)[52]. Bohr's point is that it is the wave–particle duality of radiation that makes it impossible to measure the momentum of the electron: while gamma radiation may appropriately be described in terms of the particle model, it is the indispensability of the wave model for the interpretation of the experiment that precludes the precise measurement of the momentum of the electron.

Moreover, if the wave model of the electron is employed, then the electron with a precise position, described as a narrowly confined wave-packet, cannot be said to *have* a well-defined momentum. It is, then, also the viability of the wave description of the electron that precludes the simultaneous

From duality to complementarity

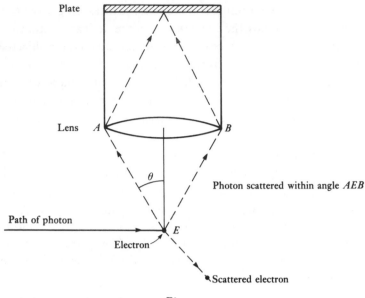

Plate

Lens A B

θ

Photon scattered within angle AEB

Path of photon

E

Electron

Scattered electron

Figure 1

precise measurement of position and momentum. The uncertainty relations are

> ... essentially an outcome of the limited accuracy with which changes in energy and momentum can be defined, when the wave-fields used for the determination of the space-time co-ordinates of the particle are sufficiently small.[53]

Bohr expounded these arguments in the paper in which his conception of complementarity was first published. He shows there that the uncertainty relations can be derived from the wave model of the electron. He notes that the Planck and de Broglie equations, $E = h\nu$ and $p = h/\lambda$, are the foundation of the corpuscular theory of radiation and the wave theory of matter. Energy and momentum are associated with the concept of a particle, while frequency and wavelength are associated with the concept of a wave. In virtue of these relations an object with a definite energy E and a definite momentum p may be described as a harmonic wave-train of frequency ν and wavelength λ. An object located at a definite position at a definite time may be described as a wave-packet formed from the superposition and mutual interference of a group of harmonic waves. The more precisely located the object, however, the more narrowly confined the wave-packet, and consequently the greater the spread of the frequencies and wavelengths of the component waves. The spread of the constituent frequencies and wavelengths of a wave-packet is given by the equations:

$$\Delta x \Delta \sigma = \Delta t \Delta \nu = 1,$$

where Δx and Δt denote the spatial and temporal extensions of the packet, and $\Delta \sigma$ and $\Delta \nu$ the spread in the wave-number and the frequency respectively. Substituting h/p for $\sigma = 1/\nu$ and E/h for ν in the above equations, the uncertainty relations, $\Delta x \Delta p_x = \Delta t \Delta E = h$, are immediately derived.[54]

Heisenberg recalls that after several weeks of discussion he and Bohr concluded that the uncertainty relations are a special case of the more general complementarity principle.[55] Heisenberg wrote to Dirac on 27 April 1927: 'Prof. Bohr says, that one in all those examples sees the very important role, which the *wave*-theory plays in my theory and, of course, he is quite right.'[56] Heisenberg added to the proofs of his article on the uncertainty relations a postscript in which he acknowledged the justice of Bohr's criticisms:

> In this connection Bohr has pointed out to me that I have overlooked essential points in some of the discussions in this work. Above all the uncertainty in the observation does not depend exclusively on the occurrence of discontinuities, but is directly connected with the necessity of doing justice simultaneously to the different experimental data which are expressed in the corpuscular theory on the one hand, and in the wave theory on the other hand.[57]

Whereas Heisenberg saw the uncertainty as a consequence of the unobservability brought about by the discontinuity, Bohr saw the unobservability as the reflection of a fundamental conceptual inadequacy. Just as matrix mechanics was the natural culmination of Bohr's correspondence approach to the development of the quantum theory, so too might the uncertainty principle be regarded as the pressing home, as it were, of Bohr's thesis that the quantum theory precludes a classical space-time description of the behaviour of microphysical objects – a view he had held at least since 1922 and which he continually emphasised.[58] Heisenberg's principle is virtually a special case of Bohr's thesis: the view that the quantum theory does not permit a classical space-time description virtually implies the thesis that the classical notion of a definite trajectory is not well defined in the quantum theory. The uncertainty principle was indeed the 'thoroughgoing revolution in the concepts upon which the description of nature has rested up to now' that Bohr had been expecting at least since July 1925.[59]

It is clear that Bohr held the uncertainty relations to be 'a direct consequence of the wave-particle dilemma'.[60] But in what way exactly?

Bohr's point is that when discontinuity obtains, an electron can no longer be conceived of as corpuscular in any well-defined sense, i.e. as a precisely localised entity. Thus he writes:

> The fundamental indeterminacy which we meet here may . . . be considered as a direct expression of the absolute limitation of the applicability of visualizable conceptions in the description of atomic phenomena, a limitation that appears in the apparent dilemma which presents itself in the question of the nature of light and of matter.[61]

The logic of Bohr's thinking here seems to be *modus tollendo tollens*: If the simultaneous position and momentum of an electron can be precisely measured, then an electron can legitimately be conceived of unequivocally as a particle; but an electron cannot be unequivocally conceived of as a particle; therefore the simultaneous position and momentum of an electron cannot be measured.

On 13 April 1927 Bohr sent Einstein a copy of the proofs of Heisenberg's paper on the uncertainty relations. The accompanying letter reveals something of Bohr's thinking at the time. He writes:

> It has of course long been known how intimately the difficulties of the quantum theory are connected with the concepts, or rather with the words, which are used in the ordinary description of nature, and which all originate in the classical theories. These concepts just give us the choice between Scylla and Charybdis accordingly as we direct our attention to the continuous or to the discontinuous side of the description. We feel at the same time, however, that the hopes conditioned by our own habits lead us into temptation here, as it has so far always been possible for us to keep swimming between the realities as long as we are prepared to sacrifice all our customary wishes. The very fact that the limitation of our concepts coincides so exactly with the limitation of our powers of observation allows us, as Heisenberg stresses, to avoid contradictions.[62]

This is rather obscure. What he means by 'Scylla' and 'Charybdis' is 'particle' and 'wave': we may conceive of an electron, say, as a particle or as a wave depending on whether continuity or discontinuity obtains. What Bohr means by the sentence 'We feel at the same time . . .' is that, provided we abandon our intuitive wish for continuity, we can avoid committing ourselves to one model or the other (i.e. we can 'swim between the realities'); that intuitive wish can be dangerous (i.e. 'may lead us into temptation'), for it can lead us to assume that matter and radiation must unequivocally be either corpuscular in nature or undulatory, an assumption that Bohr was questioning. What is meant by the last sentence is that the fact that the electron cannot be *observed* unequivocally to be a particle enables us in certain circumstances to conceive of it as a wave-like entity without blatantly contradicting any unambiguous observation.

After describing his derivation of the uncertainty relations Bohr continues:

> By means of the new formulation it is possible to harmonise the requirement of the conservation of energy with the consequences of the wave theory of light in that, according to the character of the description, the different sides of the problem never appear at the same time.[63]

The phrase 'the different sides of the problem', I think, refers back to the phrases 'the requirement of the conservation of energy' and 'the consequences of the wave theory', i.e. to the particle model and the wave model. The point is that use of the particle model is consistent with use of the wave model since the two models need not be employed with respect to the same object in one and the same experimental situation. Bohr continues:

> As for such a didactically coloured concept as visualisability it is, I think, instructive always to bear in mind how indispensable the concepts of continuous field theory are in the present state of knowledge. If we speak only of particles and quantum jumps it is hard to find a simple introduction to the theory which is based on a reference to the limitation of the possibilities of observation, since the uncertainty I spoke of depends not only on the presence of discontinuities but also on the impossibility of describing these exactly owing to those characteristics of material particles and of light which are given expression in the wave theory. The representation of an electron as a group of de Broglie waves is completely analogous to the representation of a light-quantum as a group of electro-magnetic waves.[64]

It appears from this letter and others that during the spring of 1927 Bohr's attitude to duality underwent a major change. When he learned of the uncertainty principle, he came to think that use of the wave and particle models for both radiation and matter was not just appropriate but also indispensable – indispensable in that only by employing both models could an intuitive explanation of the uncertainty principle be given. He had at last accepted wave–particle duality, though in what sense is a question I shall explore in Chapter 4. Between 1925 and 1927 his long resistance against wave-particle duality was broken down by a succession of events: the failure of the Bohr–Kramers–Slater theory, the advent of wave mechanics, the uncertainty principle, and the last straw as it were, the Davisson–Germer paper on the diffraction of electrons, which appeared at a crucial juncture, April 1927.

Consider the effect of the Davisson–Germer paper. Having insisted for so long that diffraction and interference effects necessitate use of the wave model, Bohr felt constrained to admit that if matter gives rise to diffraction effects then it too must be conceived of in terms of the wave model. Considerations such as these led him to believe that further resistance against wave-particle duality would be futile. What enabled him to reach this position without doing violence to his intellectual conscience was the insight that both models need never be applied to one and the same object at one and the same time. Both models, then, were not only necessary in some respect, but also mutually exclusive – that is, complementary.

3.5
Complementarity: summer 1927

Heisenberg recalls that Bohr developed the foundations of his conception of complementarity during his ski-ing holiday in Norway in the early part of 1927 (end of February to middle of March).[65] This recollection is correct. It seems, however, that Bohr did not hit upon the notion of (or at least the word) complementarity until the summer of that year.

The letters and manuscripts in the Niels Bohr Archive enable us to trace the development of Bohr's thinking at this time. Throughout that summer he struggled incessantly to clarify his thinking on the duality problem. Klein, who was then Bohr's assistant, recalls that Bohr dictated to him day after day that summer, each day going over the ground covered on the previous one.[66] Bohr's letters convey something of the intensity of his reflection. In a letter of 10 June to Fowler he says that he is busy preparing a paper on the more philosophical aspects of the quantum theory.[67] In a letter of 24 August to Heisenberg he writes:

> In the course of the summer I have been thinking a great deal about the general physical problems and I have, I think, succeeded to some extent in making my thoughts clear . . . These are essentially the same things we spoke so much about: but I am constantly finding new instructive points to stress, for example how the unobservability or rather the undefinability of the phase makes it possible to combine the individuality of single processes with the principle of superposition.[68]

And in a letter of 13 August 1927 to Pauli we find the term 'complementarity' – the first occurrence of the word in the correspondence:

> Thanks for your nice and most welcome letter. What you write about your and Jordan's work on electrodynamics is extremely attractive and is very much in agreement with my own view about the nature of the quantum theory, according to which the apparently contradictory requirements of superposition and individuality do not subsume contrary but complementary sides of nature.
>
> I am in complete agreement with your remarks on de Broglie's work: he is trying to achieve the impossible by a blending of the two sides of the matter.[69]

In the Bohr Archive there is a set of manuscripts and typescripts entitled 'The Philosophical Foundations of the Quantum Theory'. These papers are drafts of the work which Bohr refers to in his letter to Fowler. This work was a projected letter to *Nature* in response to the notes of N.R. Campbell and Jordan published in *Nature* on 28 May 1927.[70] Bohr's projected note to *Nature* was never published; it became instead the basis of the address which he delivered on 16 September 1927 to the International Congress of Physics at Como in commemoration of the centenary of the death of Volta. In this address Bohr expounded for the first time in public his concept of complementarity.[71]

The word 'complementary' occurs in several different drafts of Bohr's projected note to *Nature*: for example, referring to the wave–particle duality of light he writes (emendations as in the original manuscript):

> It seems that we here meet with an unavoidable dilemma, . . . the question being not of a choice between two ~~different~~ rivalising concepts but rather of the description of two complementary sides of the ~~same~~ phenomenon.[72]

A fairer copy of the above manuscript reads:

Philosophical foundations of the quantum theory
From the point of view of the principles underlying the ordinary description of physical phenomena the quantum theory represents an essentially irrational element, and we must accordingly be prepared that every concept used in accounting for experimental evidence will have only restricted validity when dealing with elementary atomic processes. This situation is brought out very clearly through the renewed discussion about the nature of light. On one hand the concept of individual light corpuscles with energy E and momentum I connected with the period of vibration σ and the wavelength λ through the quantum relations $E\sigma = I\lambda = h$ (1) may be said to be forced upon us as the rational expression for the conservation of energy and momentum in the elementary processes of interaction of light and matter. On the other hand the

appearance of the quantities σ and λ in these relations reminds us forcefully of the vast field of experience which is accounted for so conclusively by the principle of superposition of waves. It seems that we here meet with an unavoidable dilemma, the question being not a choice between two rivalling concepts but rather of the relation of two complementary sides of phenomenon [*sic*] laws of nature.[73]

In these manuscript papers Bohr frequently speaks of 'individuality' and 'superposition' rather than of 'particle' and 'wave'. By the term 'individuality' he means 'the indivisibility of elementary particles and processes'; in these energy and momentum are conserved. By 'superposition' he is referring to the ability of waves to intermingle at the same place, and also to the linearity of wave functions in wave mechanics. Thus, in what I shall call his 'Como' article he writes:

> The problem of the nature of the constituents of matter presents us with an analogous situation. The individuality of the elementary electrical corpuscles is forced upon us by general evidence. Nevertheless, recent experience, above all the discovery of the selective reflection of electrons from metal crystals, requires the use of the wave theory superposition principle in accordance with the original ideas of L. de Broglie.[74]

As we have seen, until 1926 Bohr adopted a very sceptical stance towards the question of duality: duality, he held, indicates merely the limited applicability of the classical models, and while the failure of one model makes theoretical room, so to speak, for the other model, one need not bring in the other model to fill the interpretative gap. Towards the end of 1926, however, his attitude changed somewhat: he came to hold that in any situation in which one of the two classical models fails, it is not only appropriate to employ the other but also desirable. He had hit upon one of the two main points in his thesis of complementarity, viz. that the two models are indispensable; it was the uncertainty principle that gave him the idea of the other point, viz. that the two models are applicable only in mutually exclusive experimental circumstances. Whereas in 1925 he had thought that the failure of universal continuity entailed a restriction on spatio-temporal pictures, he came to see in 1927 that what is entailed is a failure of the possibility of kinematic-dynamic description, i.e. description of the spatio-temporal and causal features of an object. We can have a well-defined spatio-temporal (kinematic) description of electrons and photons (as particles and waves respectively), and also a causal (dynamic) description of their actions and reactions against each other; but we cannot have both together.

4

The meaning
of complementarity

Bohr's general thesis of complementarity conveys the idea that we can, and perhaps must, make use of the classical physical concepts in quantum physics, notwithstanding the inadequacies of these concepts; but we can use them only within the limits circumscribed by the quantum of action – beyond these limits the classical concepts cease to be well defined. This idea originated in his thesis that the conceptual structure of the quantum theory is in a sense a generalisation of the conceptual scheme of classical mechanics; and this thesis was a corollary of the correspondence principle.

During the first months of 1927 Bohr regarded quantum mechanics as the quantum-theoretical generalisation of classical mechanics to which he had been looking forward. He came also to think that quantum mechanics required in a sense the generalisation of the two classical models of particle and wave. In classical physics these two models generally pertain to different theories, the particle model to the mechanical theory of matter, the wave model to the electromagnetic theory of radiation. In quantum physics, however, the two models pertain to one theory, quantum mechanics, and each is applicable to matter and radiation. Where discontinuity obtains, the standard classical models of matter and radiation cease to be well defined, and the alternative, non-standard models may be used. There is no logical inconsistency in this dual use of the models since each is appropriately applicable only in different empirical circumstances.

Bohr also saw the mutual exclusiveness of kinematic and dynamic concepts in quantum mechanics (such as position and time-of-occurrence, momentum and energy respectively) as a further consequence of the generalisation of classical mechanics: concepts which in the less general theory hang together, so to speak, fall apart in the more general theory. The classical notion of exact simultaneous position and momentum is, he thought, an idealisation in the sense that it is applicable, not universally, but only under special conditions: where discontinuity obtains, this notion, like the classical notion of a particle, ceases to be well defined.

Just as the laws of quantum mechanics are a generalisation of the laws of classical mechanics, so the conceptual structure of the former theory is a generalisation of the conceptual structure of the latter. The classical concepts are retained, but the logical structure governing their applicability is altered: classical predicates such as ' . . . is a particle', ' . . . has position q_x' still range over the whole domain of objects of the theory, though new restrictions are placed upon their application to objects and their combination with other predicates of the theory. Bohr puts this point by saying that the conceptual structure of classical mechanics was found to be too *narrow* to comprehend new phenomena; and it had to be *widened* by delimiting and restricting the application and combination of predicates. Thus he writes:

> . . . the very conceptual frame, appropriate both to give account of our experience in everyday life and to formulate the whole system of laws applying to the behaviour of matter in bulk . . . had to be essentially widened if it was to comprehend proper atomic phenomena.[1]

4.1
Wave-particle complementarity and kinematic-dynamic complementarity

The quantum postulate, Bohr holds, gives rise to two distinct species of complementarity. First, wave-particle complementarity:

> The two views of the nature of light are rather to be considered as different attempts at an interpretation of experimental evidence in which the limitation of the classical concepts is expressed in complementary ways. . . . In fact, here again we are not dealing with contradictory but with complementary pictures of the phenomena, which only together offer a natural generalization of the classical mode of description.[2]

Second, the complementarity of spatio-temporal descriptions and momentum-energy descriptions, which I shall call 'kinematic-dynamic' complementarity. Thus:

> The very nature of the quantum theory thus forces us to regard the space-time co-ordination and the claim of causality, the union of which characterises the classical theories, as complementary but exclusive features of the description . . .[3]

By the phrase 'claim of causality' Bohr means 'the requirement of a causal description, viz. knowledge of the energy or momentum of the object'.

As we saw in Chapter 3, Bohr seems originally to have regarded kinematic-dynamic duality as a consequence, or at least a reflexion, of wave-particle duality. He came to hold that these are simply different species or modes of complementarity, i.e. of the indispensability, yet limited applicability, of the classical concepts in the quantum theory. In his later writings he usually speaks of complementarity in terms that comprehend both modes. For example:

> Morever, evidence obtained under different conditions and rejecting comprehension in a single picture must, notwithstanding any apparent contrast, be regarded as complementary in the sense that together they exhaust all well-defined information about the atomic object.[4]

In his Como paper, after a brief introduction, Bohr states:

> The quantum theory is characterized by the acknowledgement of a fundamental limitation in the classical physical ideas when applied to atomic phenomena. The situation thus created is of a peculiar nature, since our interpretation of the experimental material rests essentially upon the classical concepts.[5]

This statement conveys the essence of Bohr's conception of complementarity: the classical physical concepts are indispensable for the theoretical interpretation of experimental data, and yet are subject to certain limitations on their use in the quantum theory. Bohr, however, does not say exactly what he means by the term 'complementary'. He talks of 'the space-time co-ordination' and 'the claim of causality' as being 'complementary but exclusive features of the description'.[6] This suggests that he understood the term in a sense close to that of its literal meaning of 'mutually or jointly completing'.[7] It is in this sense that angles and colours are said to be complementary, i.e. in that they make up a right angle and the colour white respectively.

In the Introduction to *Atomic Theory and the Description of Nature* Bohr says that the quantum of action

> ... forces us to adopt a new mode of description designated as *complementary* in the sense that any given application of classical concepts precludes the simultaneous use of other classical concepts which in a different connection are equally necessary for the elucidation of the phenomena.[8]

This sentence, the closest approach to a definition of the concept of complementarity that Bohr ever gave, suggests that the mutual exclusiveness of the applicability of the classical concepts is part of the very meaning of Bohr's notion of complementarity. Indeed, he says explicitly in the Introduction that he had used the term 'complementarity' in the Como article 'to denote the relation of mutual exclusion characteristic of the quantum theory with regard to the application of the various classical concepts and ideas'.[9] During 1929, however, he came to think that the term 'complementarity' did not effectually connote mutual exclusion at all, and that a term clearly expressing the latter idea was required. For a time he employed the term 'reciprocity' instead.[10] But when it was pointed out to him that the term 'reciprocal' does not generally signify a relation of mutual exclusion, he reverted to the use of the original term.

The literal sense of 'complementary' as 'mutually completing' captures the Bohrian notion that complementary descriptions 'represent equally essential knowledge about atomic systems and together exhaust this knowledge'.[11] There is also in the literal sense some suggestion of contrast, or of mutual exclusion even, in that complementary things frequently set each other off, as for example complements in logic and complementary colours in chromatics (such as blue and yellow).

What are complementary, strictly speaking, are concepts or propositions. Bohr, however, speaks of complementary 'concepts',[12] 'pictures',[13] 'aspects',[14] 'features',[15] 'experiences',[16] 'information',[17] 'evidence',[18] 'phenomena'.[19] Talk of complementary concepts, pictures, aspects, features is more common in Bohr's earlier writings; talk of complementary information and evidence more common in his later writings.

Two or more concepts or propositions may be said to be complementary in Bohr's sense if and only if:
(a) they are different in meaning, or predicate different properties,
(b) together, or jointly, they constitute a complete description or representation of a thing,
(c) they are mutually exclusive or incompatible either in a logical sense or in an empirical sense.
Complementary concepts generally belong to different conceptual schemes (such as kinematic and dynamic concepts) or are logically incompatible (such as the concepts of particle and wave).

Two or more descriptions of a thing are complementary only if each alone is incapable of providing a complete description or explanation of the thing in question and both together provide a complete description. In quantum mechanics the wave and particle conceptions of radiation and matter together provide a complete description or representation of the behaviour of atomic objects: each conception provides a suitable physical interpretation of some experimental findings, and together they provide interpretations of all experimental findings. Similarly, kinematic and dynamic descriptions of an object are together exhaustive, not in the sense of providing a synchronic description of the classical state of an object, but in the sense that together they provide a diachronic description comprehending the kinematic and dynamic properties of object; they jointly exhaust the possibilities of description. The main difference between the two sorts of complementarity is that, on the one hand, the concepts of particle and wave in a sense fall apart in classical physics, yet in a sense come together in quantum mechanics; and, on the other hand, the concepts of position and momentum go together in classical mechanics, yet in a sense fall apart in quantum mechanics.

Concepts and propositions which are complementary in Bohr's sense are mutually exclusive in that the application of one such concept to a certain thing at a certain time precludes the application of the other to that thing at the same time. Concepts may be mutually exclusive in a logical sense, in so far as they are logically, or conceptually, incompatible (as e.g. the concepts of wave and particle) or in an empirical or epistemic sense, in that they cannot be known through sense perception to be applicable to a thing at the same time (as for example kinematic and dynamic attributes in quantum mechanics). I shall postpone the question of whether Bohr maintained that kinematic and dynamic concepts are conceptually, as well as empirically, incompatible.

Mutual exclusiveness is frequently thought to be the sole condition of Bohr's notion of complementarity. Thus Pauli: 'When the applicability of *one* classical concept stands in an exclusive relation to that of *another*, following Bohr we call both these concepts (such as the position and momentum co-ordinates of a particle) complementary.'[20] This is a mistaken view: the notions of mutual exclusiveness and of joint completion are equally necessary, indeed complementary, ingredients in the meaning of Bohr's conception. In the genesis of the conception the notion of joint completion came first (in the acceptance of wave-particle duality); the notion of mutual exclusiveness came later (in the acceptance of the uncertainty principle).

4.2
Complementarity and consistency

In saying that the requirements of superposition and individuality (wave and particle conceptions) are not inconsistent but complementary, Bohr was purporting to resolve the ostensible paradox of duality, among other things. What, exactly, is Bohr's resolution?

Two levels of duality ought to be distinguished. Duality obtains at the formal level in that quantum mechanics, *qua* mathematical theory, may be interpreted in terms of the wave and particle conceptions. Apropos the Planck–Einstein–de Broglie relations, $E = h\nu$ and $p = h/\lambda$, Bohr writes in the Como article:

> In these formulae the two notions of light and also of matter enter in sharp contrast. While energy and momentum are associated with the concept of particles, and, hence, may be characterized according to the classical point of view by definite space-time co-ordinates, the period of vibration and wave-length refer to a plane harmonic wave train of unlimited extent in space and time.[21]

If the energy or momentum of an object is known, then a frequency or wavelength may be assigned to it, and vice versa. The apparent inconsistency in this assignment did not greatly trouble Bohr, for he regarded this sort of duality as purely formal.

Duality also obtains at the empirical level, in the physical contexts in which quantum mechanics is actually applied. Whereas 'formal' duality pertains to the interpretation of quantum mechanics *qua* formal calculus, 'empirical' duality pertains to the interpretation of empirical applications of the theory. Certain experimental evidence, Bohr held, calls only for the particle model (e.g. the Compton effect), other evidence only for the wave model (e.g. interference effects). At the empirical level one of the two models, but not both, provides a satisfactory intuitive understanding of the application of the theory. The uncertainty principle suggested to Bohr the idea that the empirical situations in which the two classical models are effectively applicable are in fact mutually exclusive. The fact that empirical duality (so to speak) is diachronic – unlike formal duality – mitigates the ostensible inconsistency.

Bohr saw in the uncertainty principle a further means of securing consistency in the interpretation of quantum mechanics.[22] The point is that in any situation in which, say, the particle model of radiation is effectively employed, it is impossible to establish by observation that the photon is a particle (i.e. an object that possesses at any instant an exact position and an exact momentum) owing to the uncertainty principle, the exact simultaneous position and momentum of the photon cannot be measured; similarly when the wave model of matter is appropriately applied, it is impossible to observe unequivocally that an electron is a wave-entity, for in that situation it is impossible to observe all of the characteristic properties of a wave: observation (highly theory-laden of course) never unequivocally establishes that a photon is a particle or an electron a wave, since the properties which define a particle are not simultaneously observable for photons, and the properties defining a wave are not simultaneously observable for electrons. The point is that when the non-standard models are called for, it cannot be empirically established that a light ray is a particle or that an electron is a wave-like entity.

In the Compton experiment for example, where the particle model of radiation is called for, the exchange of momentum between radiation and matter can be exactly determined. If radiation were corpuscular, the scattering should occur at well-defined spatial locations and not over an extended front, as it would if radiation were undulatory. But it cannot actually be observed that the scattering occurs at definite spatial points (where electrons and photons collide), for, since the momentum exchange can be accurately determined, it follows from the uncertainty principle that the position of the interaction cannot be precisely observed. If the position of the interaction could be determined at the same time, we should then have an empirically established, well-defined model of a corpuscular photon colliding with a corpuscular electron, and hence an unambiguous contradiction of the wave model of radiation. The uncertainty principle, however, rules out our decisively establishing by observation the veridical character of the corpuscular model in this case, for, if the momentum exchange is accurately determined, the position of the interaction cannot be accurately determined; the photon, as it were, cannot be caught showing all its corpuscular characteristics at once.

The position of the electron-photon interaction could of course be measured; but in that case the wave model of radiation would be more appropriate, since optical apparatus would be required for the measurement, rendering it impossible to measure the momentum exchange between electron and photon.

It might be thought that if the position of the electron at the moment of its interaction with the photon were known, the photon could be assigned this position at the moment of collision. Bohr argues that this cannot be done, since the uncertainty relations apply to the measuring instrument as well as to the measured object, in this case, to the electron as well as to the photon. If the law of conservation of momentum can be applied, as it can in this case, then the position of the electron-photon interaction cannot be established:

> . . . we see that the uncertainty in question equally affects the description of the agency of measurement and of the object.[23]

What prevents the Compton effect from being irrefragably inconsistent with the wave model of radiation is, in a sense, the wave-particle duality of the electron. The uncertainty principle leaves open, so to speak, some logical room for the wave model of radiation. This point is brought out in an unpublished lecture of 1930:

> From the conservation law we shall know the energy and the momentum with the same accuracy for the electron as for the light quantum; but that means that we shall have exactly the same latitude for both, . . . The description of the Compton effect would be impossible from the start if we did not have a corresponding dilemma for electrons as for light quanta. If we could know for sure where the electron was, we could not at all assign the interaction to the same area, could not at all maintain any sort of visualisation; . . .[24]

The same point is made elsewhere to suggest that the uncertainty principle derives from the wave-particle duality of the measuring instrument, from 'the complementary character of the pictures employed in the description of every such auxiliary agency used in the measuring process'.[25]

Just as, owing to the uncertainty principle, it is impossible to measure simultaneously all the corpuscular properties of a photon, so it is impossible to measure simultaneously all the undulatory properties of an electron. Bohr argues that the representation of an electron by a plane harmonic wave is an idealisation. For practical purposes the electron is to be represented by a wave-packet; in which case the phase cannot be established, since a wave-packet cannot be said to have a well-defined phase.[26] In a typewritten draft of the Como paper Bohr writes: 'This ambiguity in the phase, well known from the theory of optical instruments, is essential for the consistency of the representation of individuals by waves.'[27]

There is no textual evidence to suggest that Bohr held the uncertainty principle to be necessary for the resolution of the duality paradox. He says for example that the term 'complementarity' is used in atomic physics 'to characterize the relationship between experiences obtained by different experimental arrangements and visualizable only by mutually exclusive ideas . . .'[28] Wave-particle complementarity itself resolves the paradoxes of duality: since the wave and particle models are complementary, they are applicable only to mutually exclusive experimental situations.

Born, however, held that the uncertainty principle *is* required for the resolution of the duality paradox. He maintained, moreover, that it is not the case that the wave and particle models are applied only in mutually exclusive settings; each model is required for the interpretation of every phenomenon.[29] In a sense Born is right: in the Compton experiment, for example, to which the particle model is appropriate, what is actually observed (again, in a theory-laden sense) is a shift in the wavelength of the scattered radiation (wave model); a momentum is then assigned to the radiation by way of the Einstein–de Broglie equation, $p = h/\lambda$. As Bohr frequently points out, evidence for the particle model is provided only through the use of the wave model. Similarly in the Davisson–Germer experiment, what is actually measured are differences in the energies of the scattered electrons (particle model), a de Broglie wavelength being assigned via the equation, $\lambda = h/p$. Thus the application of the particle model of the photon actually requires the use of the wave model; and similarly, *mutatis mutandis*, in the case of the wave model of the electron. Nevertheless the two models are not employed at the same time or place. In the Compton experiment, whereas the wave model is applied to the passage of the scattered rays through the spectrometer, the particle model is applied to the matter-radiation interaction. A similar point holds for the Davisson–Germer experiment. Besides, the assignments of a momentum to a photon and a wavelength to an electron by means of the Planck–Einstein–de Broglie equations are instances of what I have called 'formal' duality; and complementarity has to do primarily with 'empirical' duality. Thus when the two models are applied simultaneously, they are not employed in the same sense: the one is employed in the formal sense, the other in the empirical.

The hypothesis of wave-particle complementarity, then, is taken to resolve the paradox of duality: the incompatible models need never be applied to an object at the same time; they are called for only in mutually exclusive physical situations. Bohr does not attempt to establish this hypothesis by a single argument that covers all possible cases; what he does, rather, is to argue for the hypothesis by illustrative analyses of positive instances of it, and to invite the hostile critic to suggest a counter-instance. Kinemetic-dynamic complementarity – the uncertainty principle – is not necessary for the resolution of the paradox; it operates, however, as an additional factor which makes for consistency. Thus, in situations in which the non-standard models are applied, it cannot be empirically established that an electron is an undulatory entity or radiation a corpuscular entity. If this *could* be established, wave-particle complementarity would still obtain and logical consistency would be maintained, though theoretical coherence would not, since an electron would appear radically to change its physical characteristics from moment to moment in a way that present theories could not explain. If, however, we *were* constrained to apply both models to an object at the same time, we could not unequivocally observe all its corpuscular properties or all its wave properties, owing to the uncertainty principle, and so we could not unequivocally *observe* it to be a particle or a wave.

4·3
The correlations between the two kinds of complementarity

It is sometimes stated that wave-particle complementarity is equivalent to kinematic-dynamic complementarity in that the particle model is correlated with spatio-temporal measurements or descriptions and the wave model with momentum-energy measurements. There is, however, no evidence indicating that this was Bohr's view of the correlations between the two sorts of complementarity. Léon Rosenfeld, a close colleague of Bohr's, held the opposite view: energy and momentum are associated with the particle model, position and time with the wave model.[30] Rosenfeld's view was shared by Born.[31] It may be that the latter was Bohr's own view, for he writes:

> As regards light, its propagation in space and time is adequately expressed by the electromagnetic theory . . . Nevertheless, the conservation of energy and momentum during the interaction betwen radiation and matter, as evident in the photo-electric and Compton effect, finds its adequate expression just in the light quantum idea put forward by Einstein.[32]

Which of these two views of the correlations is correct?

According to the first view the wave model is correlated with measurements of momentum or energy, for, if the momentum or energy of an object has been measured, a wavelength or frequency may be assigned by way of the Planck–Einstein–de Broglie equations. In this case it is impossible to measure the spatio-temporal location of the object, owing to the uncertainty principle. If, on the other hand, the position of an object has been measured, the object may be conceived of as a particle localised at that position; in which case, owing to the uncertainty principle, the momentum cannot be precisely measured, and consequently no well-defined wavelength can be assigned.

However plausible this first view may seem, it appears to be contradicted by the fact that (*a*) in some experiments in which energy or momentum may be measured – such as single or double slit experiments in which the diaphragms are loosely attached to the frame – the particle model is the more appropriate, since no diffraction or interference effects are observable; and (*b*) in other experiments in which position may be measured, such as slit experiments in which the diaphragms are rigidly attached to the frame, diffraction and interference effects are observable; in which case the wave model appears to be the more appropriate.

The two views of the correlation are only apparently inconsistent with each other. The first view concerns a purely formal correlation by way of the standard interpretations of the terms in the Planck–Einstein–de Broglie equations. The second view concerns an empirical correlation, as it were, in that the correlation is based on the interpretation of physical experiments in terms of models. The difference between these two views corresponds to the difference between what I have called 'formal duality' and 'empirical duality'. It is the latter sort of duality with which Bohr was primarily concerned. He writes:

> It must be rememebered that even in the indeterminacy relation (3) we are dealing with an implication of the formalism which defies unambiguous expression in words suited to describe classical physical pictures.[33]

Bohr insists that unambiguous interpretation of the quantum-mechanical formalism always requires a reference to some specific, well-defined experimental arrangement: 'the use made of the symbolic expedients will in each individual case depend upon the particular circumstances pertaining to the experimental arrangement'.[34] I do not believe, then, that there is an invariable, systematic association of one or other of the two models with either position or momentum measurement operations: wave–particle complementarity and kinematic–dynamic complementarity are logically independent notions.

4.4

| The ontological significance of wave-particle complementarity

Does Bohr hold that the two complementary models of particle and wave provide an equally realistic (or unrealistic) picture of electrons and photons?

In his earlier writings he frequently speaks of the two models as 'expressing fundamental aspects of our experience'.[35] In the manuscripts of the summer of 1927 he talks of 'complementary sides of the phenomena', 'an essential wave feature in the description of the behaviour of material particles'.[36] Vague phrases such as these might be thought to suggest that

Bohr came to hold that the wave and particle models are equally necessary for a complete description of the real nature of micro-physical entities – the symmetry thesis, as I shall call it. On this view, since the empirical evidence for the one model is just as good as the evidence for the other, electrons and photons must be pictured in terms of both models, as having the characteristics of a particle in some situations and the characteristics of a wave in other, mutually exclusive, situations. Many of Bohr's statements suggest that he held such a view. For example:

> The individuality of the elementary electrical corpuscles is forced upon us by general evidence. Nevertheless, recent experience, above all the discovery of the selective reflection of electrons from metal crystals, requires the use of the wave theory superposition principle in accordance with the original ideas of L. de Broglie.[37]

This sort of statement, however, is extremely misleading, for Bohr held no such realist position.

Bohr regarded matrix mechanics as a highly formal theory that does not readily lend itself to interpretation in terms of physical models.[38] Wave mechanics, he held, cannot be interpreted realistically in terms of matter waves. He accepted Born's probabilistic interpretation of wave mechanics from the outset: ' . . . the wave-mechanical solutions can be visualised only in so far as they can be described with the aid of the concept of free particles'.[39] Wave mechanics employs imaginary numbers; the Schrödinger wave propagates with speeds in excess of the velocity of light, and generally pertains to a $3N$-dimensional space (pertaining to a three-dimensional space only when representing a single, non-compound object). These facts indicate for Bohr 'the symbolic character of Schrödinger's method'.[40]

In the summer of 1927 he writes:

> It is a characteristic of the latest advances in the quantum theory that to an even greater extent one abandons space–time pictures in connection with the classical theories, and makes a purely symbolic use of the classical concepts . . .[41]

> I do not think that it is possible to stress the symbolic character of the quantum theoretical methods too much.[42]

What Bohr means by 'symbolic' is made clear in a letter to Christian Møller, a pupil of his who had spent the summer of 1928 studying under Schrödinger at Berlin. After a discussion with Schrödinger, Møller wrote Bohr, asking him what he meant by saying that the representation of a particle by a de

Broglie wave is 'symbolic'; Møller took this to mean that there is nothing in nature corresponding to the wave representation. In his reply Bohr states that Møller had grasped his meaning exactly:

> I have only thought to stress the fact that the circumstance that in the quantum theory we use, very largely, the same symbols as those of the classical theory does not permit us to disregard the great difference between those theories, and in particular it calls for the greatest carefulness in the use of these forms of perception to which the classical symbols are connected . . . When one thinks of the wave theory, it is, however, just its 'visualisability' which is both its strength and its weakness, and in stressing the symbolic character of the treatment I have sought to remind one of the great difference (brought about by the quantum postulate) from the classical theories, which has not always been kept sufficiently in mind.[43]

The above remarks show that Bohr did not believe that quantum mechanics admits of an unequivocally realist interpretation in terms of classical models. It is clear also that he regarded the non-standard models, the particle model of radiation and the wave model of matter, as purely formal. In a manuscript of 1929, for example, we find the notes:

> *Nature of light,* classical theory's [word illegible] and absolute limitation. Relation between wave and particle picture. The last purely symbolic. *Discovery and description of the elementary particles.* The matter waves symbolic expression for the limitation of the classical theory. The unvisualisability of the principle of superposition and the symbolic character of Schrödinger's theory $(\sqrt{-1})$ . . .[44]

But does Bohr also regard the wave model of radiation and the particle model of matter to be just as formal as the non-standard models? Indeed, no; he regards the standard classical models as having more realistic significance than the particle and wave models of radiation and matter respectively. He admits that the experimental evidence for the existence of matter waves is just as strong as the evidence for the existence of radiation waves. The concept of matter waves, however, is applicable only in circumstances in which the quantum of action must be taken into account, i.e. 'outside the domain where it is possible to carry out a causal description corresponding to our customary forms of perception and where we can ascribe to words like "the nature of matter" and "the nature of light" meanings in the ordinary sense.'[45] The point is that non-standard models are applicable only in interaction processes, where continuity, the presupposition of the meaningful applicability of such models, fails. The fertility of the wave model of matter should not be taken as indicating a 'complete analogy' between matter and radiation:

> Just as in the case of radiation quanta, . . . we have here to do with symbols helpful in the formulation of the probability laws governing the outcome of the elementary processes which cannot be further analysed in terms of classical physical ideas. In this case, phrases such as "the corpuscular nature of light" or "the wave nature of electrons" are ambiguous, since such concepts as corpuscle and wave are only well-defined within the scope of classical physics, where, of course, light and electrons are electromagnetic waves and material corpuscles respectively.[46]

In his address at the Clerk Maxwell centenary celebrations at Cambridge in October 1931 Bohr states:

> It must not be forgotten that only the classical ideas of material particles and electromagnetic waves have a field of unambiguous application, where the concepts of photons and electrons have not. Their applicability is essentially limited to cases in which, on account of the quantum of action, it is not possible to consider the phenomena observed as independent of the apparatus utilised for their observation.[47]

The standard classical models have some realistic significance, since they have a well-defined application when the presupposition of continuity holds, and also because it is these models which are applicable in the correspondence limit of high quantum numbers, where for example the electron may be regarded as a free particle and represented by a wave-packet. That this was Bohr's view is confirmed by Rosenfeld:

> Why did he give a sort of preference to Planck's view of the electromagnetic field as being in some sense more fundamental than the photon concept?
>   This is a point of view that Bohr never abandoned . . . he attached reality, i.e. reality as it was defined by him, to those aspects that could be directly observed in certain limiting circumstances, by direct macroscopic observation. And, of course, in the case of radiation it is clear that direct observation in the limiting case of small values of hv/kT gives the usual classical wave description of Hertz and Maxwell.[48]

This is corroborated by Klein, who states that Bohr made it clear that the waves employed in the interpretation of quantum mechanics were not waves in the literal, realistic sense; in the literal sense electrons are particles and radiation consists of waves.[49] Unfortunately it is not widely known that this was Bohr's view.

  This is not to say that Bohr considered the wave model of radiation and the particle model of matter as veridically realistic models *sans phrase*, but only to say that he regarded them as having some realistic significance in situations in which the precondition of the well-defined use of classical models, viz. continuity, obtains. This precondition can be regarded as being satisfied only for radiation *in vacuo* and for free material particles: radiation *in vacuo* is essentially undulatory; matter in motion is essentially corpuscular. When observation occurs, i.e. in interaction processes, this condition is generally not satisfied, and the models cease to be perfectly well defined and ought to be regarded as formal. In the limiting region of high quantum numbers, however, where discontinuity may generally be disregarded if not eliminated, the classical models may be considered as having some realistic import. The model which Bohr regards as having some realistic significance, then, is the one which is applicable in situations in which observation is approximately a classical process, i.e. in situations in which continuity can be presumed to obtain; in the case of radiation this condition is satisfied (approximately) for comparatively long wavelengths, and in the case of matter, for relatively high energies.

The standard models, however, ought not to be regarded as wholly realistic. Bohr stresses that the notions of radiation *in vacuo* and free material particles are 'abstractions', by which he means that they are hypothetical entities which cannot be observed *qua* free; they are observable only through their interaction with other objects – in which case they are no longer free.[50] These abstractions are nonetheless 'indispensable for a description of experience in connection with our ordinary space–time view' – indispensable, for example, since theories of atomic structure depend upon experiments in which atoms are bombarded with free radiation or free particles.[51] This is a critical, cautious realist conception of models.

4.5
Models and visualisability

It is clear from this account that Bohr's views on the question of the realistic import of the classical models did not undergo any radical change between 1925 and 1927. In the latter year he talks of 'a renunciation as to visualization in the ordinary sense' just as he had in the former.[52] Indeed he holds that we must expect in the future an even more thoroughgoing 'abstraction from our customary demands for a directly visualizable description of nature'.[53]

During these two years, however, Bohr gained a clearer understanding of the nature and extent of the limitations on the use of the classical models. By defining more clearly the limits of their applicability, he was able to retain a significant place for pictorial models in microphysics. Picturability breaks down only when discontinuity cannot be disregarded, i.e. in interaction processes in which the quantum of action is not negligible. It is in interaction processes that we are not justified in demanding 'a visualization by means of ordinary space-time pictures.'[54]

During 1928 and 1929 Bohr speaks constantly of the 'failure of the forms of perception adapted to our ordinary sense impressions'.[55] The expression 'forms of perception' is obviously Kant's *Anschauungsformen* (*anskuelsesformer* in Danish). For Kant the forms of perception are space and time. Bohr includes causality as well as space and time in the forms of perception: he says that 'causality may be considered as a mode of perception by which we reduce our sense impressions to order'.[56] For Kant, however, causality is a category, a concept of the understanding rather than a 'form'. Bohr's peculiar use of the phrase 'forms of perception' derives from Høffding, who talks of the 'forms of the understanding', contrary to Kant's usage. Høffding, moreover, links continuity, causality, space and time as 'forms'.[57] Bohr makes no mention of Kant in his early works, probably because he had no first-hand acquaintance with his writings.

Space, time and causality are forms of perception according to Bohr, in the sense that whatever is perceptible by means of the senses is presented in space and time, and in causal relation with other objects of perception. The forms of perception are the means by which we organise our perceptual experience: they also determine the conceptual structure of classical physics, which was designed to account for the kinematic and dynamic behaviour of macroscopic objects. For microphysical objects, however, the classical forms of perception are not necessarily applicable. In Bohr's view the breakdown of visualisability in microphysics is due not just to the submicroscopic character of the objects in question but also – and primarily – to the failure of continuity. Bohr frequently points out that the causal spatio-temporal description of classical physics presupposes continuity. It is worth noting that Høffding stressed that the forms of perception and the categories of the understanding are essentially equivalent to the requirement of continuity. One of the main themes of his philosophy in fact is a dualism of continuity and discontinuity which, he maintains, underlies almost every philosophical problem. In his *History of Modern Philosophy* for example he writes:

The law of continuity (which includes within it both the law of continuity of space and degree and the law of the causal relations of all phenomena) is valid for all phenomena, because it formulates the general conditions under which we can have real experience (as distinguished from imagination) . . . Only as the condition of experience has the law of continuity (including the causal law) validity . . .[58]

This statement of Høffding's enables us to understand what Bohr meant when he said in 1926–27 that 'the definition of every word essentially presupposes the continuity of phenomena and becomes ambiguous as soon as this presupposition no longer applies'.[59] On this view, ordinary language, which is designed for the organisation of our ordinary sensory experience, presupposes continuity; hence, when this presupposition fails, ordinary language, including the models in terms of which we interpret physical theory, becomes not-well-defined, or 'ambiguous' as Bohr puts it.

Furthermore, Høffding regards the forms of perception as 'abstractions and ideals': they have an 'ideal perfection' to which nothing in our sensory experience exactly corresponds.[60] Bohr also regards the forms of perception and the classical physical concepts as abstractions and idealisations in this sense, and also in the technical sense that they presuppose extreme or limiting values of certain physical quantities. After 1927 Bohr emphasises what he saw as a profound similarity between the quantum theory and the theory of relativity:

In both cases we are concerned with the recognition of physical laws which lie outside the domain of our ordinary experience and which present difficulties to our accustomed forms of perception. We learn that these forms of perception are *idealizations*, the suitability of which for reducing our ordinary sense impressions to order depends upon the practically infinite velocity of light and upon the smallness of the quantum of action.[61]

Just as our sharp distinction between space and time depends upon the small value of the velocities of the objects we perceive, so our usual causal, spatio-temporal description of objects depends upon the relatively large size of the objects we observe. In spite of their limitation the forms of perception are nevertheless indispensable: they 'colour our whole language' and 'all experience must ultimately be expressed' in terms of them.[62] Consequently it is misconceived to think that the problems of atomic theory can be evaded by 'replacing the concepts of classical physics by new conceptual forms'.[63] The Kantian character of this view is unmistakable: the forms of perception are the preconditions of the possibility of our sensory experience and of the meanings of the words we use to describe it, including the theoretical elaborations of ordinary language employed in physics. It is clear why Bohr rejects a realist interpretation of wave-particle complementarity: when the quantum of action cannot properly be ignored, the classical models cease to have any well-defined applicability, and expressions like 'the nature of matter' and 'the nature of light' cease to have any well-defined meaning.[64] I shall discuss the Kantian aspects of Bohr's philosophy in more detail in Chapter 10.

4.6
Bohr's view of models

Before discussing Bohr's view of models in greater detail it will be useful to survey very briefly the main roles which models are commonly assigned in physics.

Models, in the sense in which I am using this term, have two main functions in physics: they may be proposed either as putatively true representations of the physical characteristics of the objects treated by some theory, or as purely imaginary devices, heuristic fictions.

Firstly, models may play a heuristic role, in suggesting a set of fruitful concepts or mathematical methods for some theory. Clerk Maxwell, for example, devised his electrostatic theory by conceiving of an electric field as an incompressible, frictionless fluid flowing through tubes, the flow satisfying the same partial differential equations as those in hydrodynamics.[65] Again, Bohr's planetary model of the atom performed this role remarkably well. In these examples the model plays an important part in the formation of the theory: models, to use Clerk Maxwell's happy phrase, may be 'science forming'. Heuristic models may also be useful in suggesting ways in which a theory, once formed, may be elaborated, extended or modified in order to account for fresh or recalcitrant empirical data. J.D. van der Waals, for example, used the molecular model of a gas to generalise the molecular theory of heat: he took into account the radii of the molecules and the forces acting between molecules, thus tapping the latent or untapped analogy of the model. Goudsmit and Uhlenbeck did something similar in their hypothesis of electron spin. Successful heuristic models increase the explanatory and predictive power of a theory. If a heuristic model is not taken to provide a true representation of the characteristics of the intended objects of the theory it is what I have called a 'formal' model. Clerk Maxwell's hydrodynamic model of elecrostatics, for example, was purely formal: he regarded the fluid as purely imaginary.

Secondly, models may have the function of providing putatively true, visualisable descriptions of the intended objects of a theory. When first proposed, a model is generally treated as a heuristic fiction; if it proves very successful, however, the question arises as to whether, and to what extent, it may provide a true representation of the objects of the theory. Theoretical objects, which may originally have been proposed as purely hypothetical entities, may, on the success of the model, come to be treated as real entities, or as plausible candidates for the title of real entity. Thus the models of atoms and molecules came to be treated as realistic. Towards the end of the nineteenth century the atom was generally regarded as a hypothetical entity; it came to be treated as a real entity during the first two decades of the twentieth century, when the work of Einstein and Jean Perrin on Brownian motion, and of Rutherford and others on radio-activity, brilliantly demonstrated the heuristic power of the model.

It is generally agreed that formal models play a very useful role in physics, though not that they are actually necessary for the formation and development of theories. Those who hold a non-realist theory of science deny that realistic models have any genuine role to play in physics; in their view the aim of a physical theory is to 'save' the phenomena, i.e. to devise deterministic or statistical laws which may be used to cover past observable appearances and predict future ones. Those who adopt a realist philosophy of science maintain that a physical theory is required to do more than merely 'save' the phenomena: it is required to account for the covering laws in terms of the characteristics of the real, though unobservable, entities which give rise to the observable appearances, for only in this way are the phenomena genuinely explained. To this end models are essential ingredients in physical theory, and not merely auxiliary trappings.

Bohr made considerable use of models in his physics: much of his most important work was based on them, as for example his theories of atomic structure of 1913 and 1922, his theory of the compound nucleus of 1936, and his liquid-drop theory of nuclear structure in 1939. Although he employed models to great effect, he always treated them with circumspection. Yet up until 1920 he seems to have been fairly sanguine about the realistic content of his models of atomic structure. To N.R. Campbell's suggestion that Bohr's planetary model of the atom is purely formal, that the electrons within the atom do not orbit the nucleus, he replied that the fact that his theory of the atom provided an explanation of atomic spectra appeared to him to be an argument in favour of the reality of his assumptions.[66] When this statement was made, his new theory of the shell structure of the atom was achieving brilliant success in the explanation of the periodic table of the elements. As the years went by, however, his confidence in the realistic import of his models waned.

We can gain some idea of Bohr's view of models in the early twenties from a letter to Høffding. The notion of analogy is a *leitmotiv* in Høffding's writings: in his *Problems of Philosophy* he states that analogy is of great importance for description and discovery in science and that the atomic theory and the mechanical conception of nature are 'so many vast analogies'.[67] In his letter to Høffding Bohr writes:

> The question of the role of analogy in scientific investigations which you stressed is undoubtedly an essential feature of every study in the natural sciences, even if it does not always stand out. It is often quite possible to make use of a picture of a geometrical or arithmetical sort which covers the problem in question in such a clear way that the considerations almost acquire a purely logical character. In general, and particularly in some new fields of research, one must however constantly keep in mind the obvious or possible inadequacy of the picture, and, so long as the analogies make a strong showing, be content if the usefulness or rather fruitfulness in the area they are used is beyond doubt. Such a state of affairs holds not least from the standpoint of the present atomic theory. Here we are in the peculiar situation that we have gained some information about the structure of the atom which may safely be considered just as certain as any of the facts of natural science. On the other hand we meet with difficulties of such a profound nature that we cannot see any way of solving them; in my personal opinion these difficulties are of such a kind that they scarcely allow us any hope of carrying through in the atomic realm a description in time and space of the kind that matches our ordinary sense impressions. In these circumstances one must naturally bear in mind that one is operating with analogies, and the point, that the areas of use of these analogies in the individual case are restricted, is of decisive importance for progress.[68]

This letter was written at a time when Bohr's models, successful in many ways though they were, were facing serious difficulties. It is clear, however, that his attitude to models is extremely cautious. He believes that the models of atomic structure have some realistic significance (cf. ' . . . we have gained some information about the structure of the atom . . .'), but he is acutely conscious of the negative analogy of the models; indeed he doubts whether a complete, realistic model of atomic processes is obtainable. It does not seem from this letter that Bohr believed that the models were wholly formal; nor does it convey a robust realism, but at the most, a diffident, circumspect realism. Pauli recalls that while in Copenhagen in 1922–23 he was impressed by the cautiousness with which Bohr, in contrast to other physicists, employed classical models.[69] Pauli was himself a severe critic of the use of models in atomic physics: he completely rejected the notion of electron orbits.[70] His own anti-realism was inspired by his commitment to Machian positivism.

Bohr's attitude to the classical models did not change very much after the early twenties. In 1929 he stated that the concept of stationary states of the atom may be said to possess, within its field of application, just as much, or if one prefers, just as little reality as the notion of an elementary particle. These notions are expedients which are helpful in expressing in a consistent manner 'essential aspects of the phenomena'.[71] He maintained, nevertheless, that every doubt regarding the reality of atoms had been removed, and that we have gained a detailed knowledge even of the inner structure of the atom.[72]

In Bohr's view, then, models help us to construct theories which enable us to explain and predict the course of our sensory experience. Highly successful models may owe their success to the fact that they faithfully represent at least some aspects of the real entities which lie beneath the appearances. Since, however, there are limits to our attaining wholly realistic models in microphysics, we ought not to put too much faith in the realistic performance of models: fulfilment of the realistic role is a benefit devoutly to be wished for, but not confidently to be expected.

I wish now to take a critical look at the idea of wave-particle complementarity, before going on in the next chapter to examine the notion of kinematic-dynamic complementarity.

4.7
A critique of wave–particle complementarity

Bohr often states that the classical models of particle and wave are equally necessary for the theory of matter and the theory of radiation.[73] This thesis – the symmetry thesis, as I have called it – expresses one of the main points in calling the two models 'complementary'. The thesis, however, is problematical: it çannot be understood in realist terms, for, as we have seen, Bohr, does not assign any realistic significance to the non-standard models of matter and radiation. The two models, then, cannot be said to be equally necessary, or at least necessary in the same respect.

Why, then, are the non-standard models needed at all? Bohr sometimes suggests that they enable us to construct visualisable pictures of those aspects of matter and radiation that cannot be understood in terms of the standard models. In the manuscripts of 1927 we find the following statements:

> The theory therefore does not make room for any directly visualisable element, but a 'visualisability' can be achieved by keeping the complementary sides of the question before one's eyes.[74]
> In fact in our attempt to visualise (to make ourselves at home) the wave concept for matter is just as essential as it is in optics . . .[75]

This suggests that the non-standard models are mere sops to our habitual desire for pictorial representations. These models, then, are conditionally necessary, necessary given the desire for visualisation. But ought we not to resist this desire, if avowedly it cannot be satisfied by pictures which faithfully portray the real properties of the objects? Indeed, yes; but in that case the necessity of the non-standard models, and with it the symmetry thesis, dissolves.

Bohr sometimes suggests that the non-standard models are essential for heuristic reasons: they are 'invaluable expedients' for the formulation of the statistical laws governing such phenomena as the photo-electric effect and electron diffraction.[76] But have the non-standard models not long since exhausted their heuristic usefulness? Doubtless they *were* heuristically useful between 1905 and 1927, but are they *still* useful in that respect? This is doubtful, to say the least: it is plausible to suggest that what now matters in quantum mechanics are the theoretical laws themselves and not the fictive, pictorial devices which were instrumental in the invention of these laws; once we have the laws, the models may be discarded as beautiful, though now redundant, chrysalises.

One could, of course, construe the symmetry thesis in instrumentalist terms, treating the standard and the non-standard models as purely formal devices for predicting the results of experiments. On this reading the two sorts of models would be equally necessary for 'saving the phenomena'; they would relate to different *phenomena* in Bohr's technical sense of 'the object as it appears in the context of an entire experimental arrangement'. Since the two models relate merely to appearances and not to reality, one ought not to ask which model gives a truer representation of the object as it is in itself. This construal, however, is not open to Bohr, since he does not construe the standard models in instrumentalist terms.

The symmetry thesis, then, is difficult to sustain, and with it the thesis of wave-particle complementarity. The thesis has lost the palliative value it once had, and has now merely a historical significance. When it was introduced in 1927, it was a useful stop-gap measure which enabled physicists to say 'Pending the invention of a theory that is susceptible of interpretation in terms of single, unified models, let us not worry too much about the problem of duality; let us use both models as the situation demands'. Bearing in mind Bohr's unique perspective on this problem, we can understand how he was inclined to regard complementarity as having more than a purely temporary significance, as expressing rather an indelible feature of the conceptual structure of physics. Having resisted wave-particle duality with heroic tenacity for so long, he embraced duality with, as it were, rebound intensity when his resistance to it ultimately snapped. Complementarity was the device which Bohr employed to assuage his intellectual conscience when at last he gave in to the idea of duality. In his more careful moments, however, Bohr recognised that the conceptions of the electron as a matter-wave and of the light-quantum as an electromagnetic particle are simply picturesque metaphorical expressions of the fact that electrons are not *classical* particles, entities with an exact simultaneous position and momentum, and that light is not a *classical* wave-entity, an entity that is spread out over a wide front. In recognising this, however, he was cutting the ground from under the feet of the notion of wave-particle duality. Perhaps that was why he was not eager to broadcast his view that the non-standard models have no realistic import. While the symmetry thesis is difficult for Bohr to sustain, the thesis of mutual exclusiveness remains intact. The latter thesis, however, is not on its own sufficient to sustain the thesis of wave-particle complementarity.

But what of the notion of kinematic-dynamic complementarity? This, I believe, is the much more interesting and important of the two notions of complementarity. I shall examine it in the following chapter.

5
The foundations of
kinematic–dynamic complementarity

Having discussed the meaning of complementarity in general and wave-particle complementarity in particular, I wish now to explicate Bohr's subtle argument for the thesis that certain pairs of classical concepts which are mutually compatible in classical physics are mutually exclusive in quantum physics, i.e. the thesis of kinematic–dynamic complementarity. It is generally held that Bohr regarded the mutual exclusiveness in question as being not only epistemic but also ontic or semantic, in the sense that the uncertainty principle expresses a restriction not only on the joint measurability of canonically conjugate observables but also on the joint instantiation of the properties corresponding to these observables. I shall consider the question of the ontic or semantic construal of kinematic–dynamic complementarity in Chapter 7. Bohr's argument for complementarity in the epistemic sense is, however, more fundamental: it goes to the heart of his theory of observation and measurement in quantum physics. What, then, is Bohr's argument?

5.1
The mutual exclusiveness of kinematic and dynamic properties

Kinematic and dynamic attributes in quantum mechanics are mutually exclusive in the sense that they cannot be simultaneously measured; they are, in this sense, epistemically incompatible. Why did Bohr think this? His reasons may seem intricate, but they are basically quite simple. Two main factors, he maintains, account for the epistemic incompatibility: (*a*) measurements of these two different sorts of properties require mutually exclusive experimental arrangements, incompatible measurement procedures; (*b*) the indeterminability of the interaction between the object and the instrument of measurement precludes extrapolation of the different measurement results to one and the same time. Bohr argues for this thesis of epistemic incompatibility by analysing various experimental arrangements.

The position of a microphysical object could in principle be measured by employing a 'gamma-ray microscope'. In this case the accuracy of the measurement would be determined by the angular aperture of the lens of the microscope together with the wavelength of the radiation: thus,

$$\Delta x \sim \lambda/2\sin\theta,$$

where 'Δx' is the uncertainty in the position of the object, 'λ' the wavelength of the radiation passing through the microscope, and 'θ' is half the angle of aperture. The smaller we wished Δx to be, the shorter the wavelength of the radiation would have to be. Moreover, there would be an uncertainty Δt in the time t at which the position of the object was measured, an uncertainty which could not be less than $1/\nu$ where ν is the average frequency of the radiation; though Δt could be made arbitrarily small by using radiation of arbitrarily short wavelength. However, since the direction of the scattered photon could not be determined within the angle 2θ, the momentum of the object in the x-direction after the measurement would be uncertain by the amount,

$$\Delta p_x \sim 2\theta h/\lambda \text{ (see Figure 1)}.$$

The momentum of an object could in principle be determined with arbitrary accuracy by measuring the Doppler shift (or Compton shift) of radiation scattered from it. The disturbance of the object's position could be minimised by employing radiation of very long wavelength, though it could not be eliminated completely. The accuracy of the measurement of the change in wavelength of the scattered radiation would depend upon the length of the wave-train. Assuming that the directions of the incident and scattered radiation were respectively parallel and opposite to the direction of motion of the object, then $c\lambda/2L$ could be taken as a measure of the accuracy in the measurement of the velocity of the object ('L' signifying the length of the wave-train), the corresponding uncertainty in the momentum of the object being $cm\lambda/2L$ ('m' denoting the mass of the object). The longer the wave-train, and consequently the longer the duration of the measurement process, the more accurate the measurement; hence the position of the object during the scattering process would be uncertain by an amount $2hL/mc\lambda$.

We might think that we could measure precisely the simultaneous position and momentum of an object by illuminating it simultaneously with gamma radiation and light of long wavelength. These two sorts of measurement, however, could not be combined to provide an arbitrarily precise measurement of the simultaneous position and momentum of the object, since the time required for a comparatively precise measurement of position would be very much shorter than the time required for a relatively precise measurement of momentum; and besides, the momentum of the object would be disturbed by the position measurement during the latter time interval. Consequently, simultaneity could not be achieved.

The momentum of an object could in principle be measured by the time-of-flight method if the mass of the object were already known. If two successive measurements of the position of the object were made, separated by a short time interval $\sim t$, then the average momentum during Δt would be $m\Delta x/\Delta t$, the instantaneous momentum being $\lim_{t>0} m\Delta x/\Delta t$. As Δx decreased, however, any uncertainty in the position measurement would increasingly affect the accuracy of the momentum measurement. High accuracy in the position measurement would require use of radiation of very high frequency, entailing a large and indeterminable disturbance of the momentum of the object; and so the momentum of the object before the first position measurement at t_0 would be different from its momentum after t_0, and different again after the second measurement at t_n. Consequently, measurement of the momentum of the object in this way would tell us nothing about its momentum immediately before t_0 and immediately after t_n. What we had measured would be a quantity created by the "measurement" process, rather than a property possessed by the object independently of the measurement process. This is what Bohr has in mind when he writes:

> Indeed, the position of an individual at two given moments can be measured with any desired degree of accuracy; but if, from such measurements, we would calculate the velocity of the individual in the ordinary way, it must be clearly realized that we are dealing with an abstraction, from which no unambiguous information concerning the previous or future behaviour of the individual can be obtained.[1]

By 'abstraction' here he means that the measured velocity would be an artefact created by the process of "measurement".

J.L. Park and Henry Margenau have proposed an example of joint measurement which is less of an abstraction in the sense that the measured momentum is not an artefact of the process of measurement. They consider an electron prepared in the state $\psi(x, t = 0)$ of compact support which moves freely along the x-co-ordinate axis, with the origin arranged so that the interval where $\psi(x) \neq 0$ is $(-x_0, x_0)$. If after a very long time $t \to \infty$ the position of the electron were measured at x_n, then the momentum $mx_{n/t}$ could be calculated with great accuracy. Assuming that the momentum of the electron was constant during the interval (t_0, t_n), the average momentum determined in this way could be ascribed to time t_0.[2] This example shows, Park and Margenau maintain, that contrary to what is generally believed, it is not universally the case that for the two observables A, B it is possible to determine the value of A as a function of B only if the operators corresponding to the observables commute. They add that, since the quantum-mechanical density operator does not provide joint probability distributions for the values of non-commuting operators, the joint values obtained in this example (which they think must exist) constitute a theoretical anomaly which is not explained by the quantum theory of single measurements.[3]

Examples such as these, I believe, should on the face of it be regarded as genuine counter-examples to the uncertainty principle understood as the thesis to the effect that the simultaneous exact position and exact momentum of an object cannot be *known*. But are they genuine counter-examples to the uncertainty principle as Bohr understood it? I think not, since Bohr generally, though not always, expresses the uncertainty principle as a thesis concerning limitations on the *measurability*, as distinct from the *knowability*, of certain physical properties. Thus,

> ... Heisenberg has given the relation (2) as an expression for the maximum precision with which the space-time co-ordinates and momentum-energy components of a particle can be measured simultaneously.[4]

Besides, Bohr generally explains the uncertainty principle, or kinematic-dynamic complementarity, by analysing specific measurement operations. These analyses make it clear that what are mutually exclusive are space-time measurement operations and momentum-energy measurement operations. It is important to note here that in Park and Margenau's example there is only one measurement operation, not two; indeed they make it clear that they are concerned only with what they call 'trivial joint measurement', i.e. the calculation of the value of some observable A on the basis of the measured value of another observable B which is a function of A (i.e. $B = f(A)$).[5] As Bohr understands it, however, the uncertainty principle has to do with what he calls 'direct measurements':

> ... the information about the state of an atomic system, obtainable by direct measurements, must always imply a reciprocal latitude as regards the values of any pair of canonically conjugate variables ...[6]

Although Park and Margenau's example leaves intact Bohr's epistemic construal of complementarity, it does, however, present a difficulty for the ontic construal. I shall consider the ontic construal in Chapter 7.

To return to Bohr's analysis of measurement operations: he points out that it is only in classical physics that momentum and energy can be precisely determined on the basis of spatio-temporal measurements; in quantum physics determination of these properties usually requires employment of the conservation laws. Only in the classical limit of quantum mechanics can energy and momentum be unambiguously defined on the basis of spatio-temporal pictures.[7]

Bohr argues that if the position of an object is to be measured with a precision within a range in which the uncertainty relations become empirically significant, then the position of the measuring instrument must be known with a comparable exactness. This condition is satisfied, he argues, only if the measuring instrument is fixed rigidly to the apparatus defining the spatial reference frame. This condition, however, is incompatible with the necessary condition for the measurement of exact momentum, viz. that the measuring instrument is not rigidly attached to the structure defining the spatial reference frame. In order to determine the momentum of the object by employing the law of the conservation of momentum, the prior momentum of the instrument must be known, and the change in its momentum due to its interaction with the object must be measurable. But this change can be measured only if the instrument is not rigidly attached to the apparatus defining the spatial reference frame: the 'looseness' of the measuring instrument entails that its position is to some degree uncertain. Thus the experimental conditions required for the simultaneous measurement of exact position and exact momentum are mutually exclusive.

Similar considerations hold good concerning the measurement of time and energy. Energy measurements are generally similar to momentum measurements.[8] Measurement of time requires a clock which is precisely synchronised with the process defining the temporal reference frame, and which is constructed in such a way that it is not appreciably affected by the process or event being timed or by the reading of the time registered. This being the case, the clock is incapable of functioning as an energy-measuring instrument: the very imperturbability of the time-measuring device renders it unsuitable for measuring energy, since an energy-measuring device *eo ipso* must be perturbable. Bohr constantly stresses that spatio-temporal measurements require a co-ordinate system fixed by rigid bodies and imperturbable clocks.[9] Precise spatio-temporal measurements require a transfer of momentum and energy to fixed scales and synchronised clocks which cannot be determined or controlled if these instruments are to fulfil the role of defining the reference frame.[10] Precise momentum and energy measurements, by contrast, require a loose connection between the measuring instruments and the objects defining the spatio-temporal reference frame.

Bohr's thesis that position measurements and momentum measurements require respectively a loose, and a rigid, connection between the object and the measuring instrument is of profound importance in his epistemological analysis of quantum physics. This thesis alone, however, is not sufficient to account for the uncertainty principle. The mutual exclusiveness of measurement operations would be of little consequence if position and time measurements involved only a negligible, or a determinable, interaction between the object and the measuring instrument, since extrapolation of the results to other times would then be possible. If, say, one measured the position of an object at time t_0 and its momentum at a later time t_n, the result of the second measurement could be assigned to the earlier time, provided the disturbance of the state of the object caused by the first measurement could be ascertained and taken into account. If it could not, then the physical state of the object would not be the same at the different times t_0 and t_n; and so the extrapolation could not be made.[11] Thus the interaction between the object and the measuring instrument is indeterminable. Bohr writes:

> In particular, the impossibility of a separate control of the interaction between the atomic objects and the instruments indispensable for the definition of the experimental conditions prevents the unrestricted combination of space-time coordination and dynamical conservation laws on which the deterministic description in classical physics rests.[12]

5.2
The indeterminability of the measurement interaction

In his Como paper Bohr states that observation and measurement rest ultimately on the spatio-temporal coincidence of two independent events. All observation involves an interaction between two objects, the object of observation and the instrument of observation. The interaction is a finite exchange of energy and momentum between object and instrument. It is, however, a presupposition of classical physics that 'the phenomena concerned may be observed without disturbing them appreciably'.[13] The measurement interaction, according to classical physics, may in principle be made arbitrarily small, and it is in general either negligible, or determinable and controllable. Frequently the interaction is negligible, the total energy of the observed object being immensely greater than the energy exchanged in the interaction; if it is not negligible, it is generally determinable, i.e. capable of being accurately estimated. In quantum physics, however, the measurement interaction is in general not negligible, since the energy exchanged is large relative to the total energy of the object, and cannot be made arbitrarily small, owing to the quantum of action. But most important of all, the interaction involved in *position* measurements is indeterminable. The interaction between object and instrument which is involved in a position measurement is indeterminable in virtue of the fact that the instrument must be rigidly connected to the macroscopic apparatus defining the spatial reference frame: the energy and momentum which the instrument gains or loses in the measurement process disappear irretrievably within the surrounding apparatus. Thus, if on the one hand we *observe* an object, i.e. determine its spatio-temporal location, we interfere with its dynamic state; if on the other hand we determine its dynamic state, we are debarred from observing its position.[14]

There are, then, two factors which account for the mutual exclusiveness of exact simultaneous measurements of position and momentum. First, the mutual exclusiveness of the measurement arrangements: the one measurement requires a fixed instrument, the other a loose instrument. This factor is not wholly independent of quantum considerations, since it depends on the fact that, owing to the disturbance of the object entailed by the measuring process, measurement of the momentum of an object requires application of the conservation laws. The second factor is the indeterminable disturbance of the object; and this is a consequence of the quantum of action. Each factor alone would not give rise to mutual exclusiveness; but together they do. In this way Bohr explains the precise role which the quantum of action plays in the uncertainty principle – in kinematic-dynamic complementarity.

The mutual exclusiveness of measurement operations is, Bohr argues, reflected in the commutation relation,

$$pq - qp = \hbar i,$$

which can be taken to signify that the result of a momentum measurement followed by a position measurement is different from the result of a position measurement followed by a momentum measurement, the quantum of action being a measure of the difference. He writes:

> Indeed, it became evident that the formal representation of physical quantities by non-commuting operators directly reflects the relationship of mutual exclusion between the operations by which the respective physical quantities are defined and measured.[15]

5.3
The distinction between object and instrument

It might be thought that the momentum exchange involved in a position measurement could be determined if the momentum of the instrument were measured before and after its interaction with the object: owing to the conservation of momentum, the momentum exchanged would be the difference between the momentum of the object before and after the interaction. Bohr argues that this determination cannot be made if the instrument is to serve its function of measuring the exact position of the object. For, in order for the determination to be made, the instrument would have to be treated as an object; in which case it too would be subject to the uncertainty relations. If the momentum of the instrument at the time of the position measurement were known with an accuracy sufficient for determining the momentum exchanged, then, in view of the uncertainty relations, its position could not be accurately known; in which case it could not be used to measure the exact position of the object. By invoking the uncertainty relations Bohr may seem to be arguing in a circle here. There is, however, no vicious circularity: the essential point of the argument is that if the momentum of the position-measuring device is to be precisely determined before and after the position measurement, then *ipso facto* it ceases to be part of the system defining the spatial reference frame, and hence cannot be used to measure precise position. Bohr's point is that in order to determine the exchange of momentum between the object and the instrument, the latter would have to be treated as an object; in which case, owing to the mutual exclusiveness of position and momentum measurement operations, we could not determine both the position and the momentum of the instrument. By treating the measuring instrument as being itself an object one would simply be pushing the problem one stage further back, and one could go on doing this *ad infinitum*, without ever resolving the problem.

Bohr illustrates this point in the Como paper by means of the imaginary 'gamma-ray microscope' experiment. If the position of an electron, say, were measured with a 'gamma-ray microscope', it might be thought that the exchange of momentum involved in the measurement could be determined by measuring the change in the momentum of the microscope, including the radiation source and the photographic plate. This, however, could not be done if the position of the microscope were to be determined with an exactness relative to the spatial reference frame that was sufficient to enable it to perform a precise measurement of the position of the particle.[16]

Consider also the single-slit experiment. The y-co-ordinate of an object could be determined by passing the object through a diaphragm containing a slit, the precision of the measurement being determined by the height d of the slit ($\Delta y = d$). For this measurement the diaphragm would have to be rigidly attached to the support defining the spatial reference frame. Owing to the rigid connection between the diaphragm and the rest of the apparatus, however, the momentum exchanged between object and diaphragm would disappear irretrievably into the apparatus (see Figure 2).[17]

The momentum exchanged could be determined if the momentum of the object before its traversal of the slit, and the momentum of the diaphragm before and after traversal, were known, for the law of conservation of momentum could then be applied to the combined system of object and diaphragm. Measurement of the momentum of the diaphragm, however, would be possible only if the diaphragm were loosely attached to the rest of the apparatus and allowed to move freely, for example by being suspended

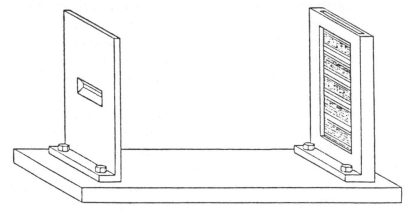

Figure 2
(By courtesy of Prof. P.E. Schilpp.)

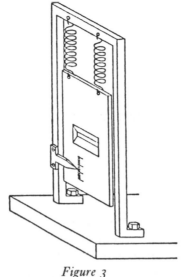

Figure 3
(By courtesy of Prof. P.E. Schilpp.)

from a weak spring. The loose connection, however, would preclude the diaphragm's functioning as an instrument for measuring position, since its position during the interval between the two momentum measurements would be uncertain by an amount determined by the uncertainty relations (see Figure 3). Any attempt to survey the position measurement interaction would necessitate the diaphragm's being treated as an *object* of measurement, as part of the system being observed; in which case the uncertainty relations could not be ignored with respect to it, and it could no longer function as an instrument of measurement. The uncertainty $\Delta p_y'$ in the momentum of the object after passing through the diaphragm is $\Delta p_y' = p_y \sin A$, where 'p_y' is the momentum of the object before its passage through the slit, and 'A' the angle between the x-axis and the first diffraction minimum. In order that the momentum of the object after passage through the diaphragm be accurately known, the momentum of the diaphragm after passage would have to be measured with an uncertainty $\delta p_y < p_y \sin A$. Since $p_y \sin A$ equals $h/\lambda \times \lambda/2d$, we have $\delta p_y < h/2d$. Taking the uncertainty relations for the diaphragm into account, $(\delta y \, \delta p_y \geqslant h/2)$, we have $\delta y > d$. Thus the uncertainty in the position of the diaphragm would be of an amount exceeding the height of the slit (see Figure 4).[18] Bohr writes:

> ... any attempt at an ordering in space–time leads to a break in the causal chain, since such an attempt is bound up with an essential exchange of momentum

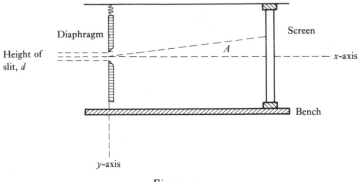

Figure 4

and energy between the individuals and the measuring rods and clocks used for observation; and just this exchange cannot be taken into account if the measuring instruments are to fulfil their purpose.[19]

If the diaphragm were rigidly connected to the rest of the apparatus, the momentum conveyed to the whole apparatus by the object could in principle be determined in a similar way, by measuring the momentum of the whole apparatus before and after the interaction. But, again, the position of the apparatus with respect to the spatial reference frame of the laboratory would be uncertain; and so we would not know exactly where in the laboratory the object was when it passed through the diaphragm. Thus, we may either regard the diaphragm as an *instrument*, as part of the measuring apparatus, in which case we may disregard the uncertainty relations with respect to it, thereby forfeiting the opportunity to determine the extent of the interaction with the object; or we may treat the diaphragm as an *object* of measurement, in which case we must take into account the uncertainty relations with respect to it, thereby losing the opportunity to perform the measurement originally intended; but we may not do both. The measurement interaction, then, can be determined only if the measuring instrument ceases to function as an instrument, that is, only if its well-defined connection with the spatial reference frame is lost.

5·4
Wholeness: the integrity of the conditions of observation

Owing to the inevitable interaction involved in measurement there can, Bohr argues, be no sharp dynamical distinction between the object and the measuring instrument:

. . . the finite magnitude of the quantum of action prevents altogether a sharp distinction being made between a phenomenon and the agency by which it is observed, a distinction which underlies the customary concept of observation and, therefore, forms the basis of the classical ideas of motion.[20]

During the process of observation the object and the instrument cannot be regarded as completely separate entities: they are joined together, as it were, by the interaction between them. An isolated, unobserved object, such as an atom in a definite stationary state, which is not interacting with any other object, is a 'closed' system. Observation renders the object an 'open' system which is not in any definite dynamic state during the observation process. The combined system of object in interaction with the instrument constitutes itself a sort of closed system.[21] Any attempt to analyse this system, to separate the object sharply from the instrument by surveying the observation interaction, would violate its essential *wholeness*. Just as the quantum of action is itself a 'feature of wholeness',[22] an element of 'individuality'[23] (by which Bohr means 'indivisibility'), so too each quantum process, and the objects involved in it, has an essential wholeness or indivisibility.

The essential wholeness constituted by object and instrument in interaction with each other precludes any analysis of the interaction, since any attempt to 'subdivide the phenomenon' would give rise to a new, complementary phenomenon:

> The essential wholeness of a proper quantum phenomenon finds indeed logical expression in the circumstance that any attempt at its well-defined subdivision would require a change in the experimental arrangement incompatible with the appearance of the phenomemon itself.[24]

Bohr's point is that surveyal of the object-instrument interaction would be possible only if the instrument were suitably modified, or if some further measuring device were introduced, i.e. only if the original instrument ceased to function as an instrument and became itself an object of investigation. Modification of the original experimental arrangement would destroy the original conditions of observation; introducing a further measuring device would give rise to different results, to 'new individual phenomena'.[25] The observation interaction, then, is not unconditionally indeterminable; it is indeterminable only on the condition that the intended measurement shall be made.

The two-slit experiment provides a striking example of Bohr's thesis of wholeness and the integrity of the conditions of observation. If the diaphragm is fixed rigidly to the system defining the spatio-temporal reference frame, it is impossible to determine through which slit an object passes without destroying the interference effect upon the screen. Various ways have been suggested for determining through which slit an object passes: illuminating the right-hand side of the diaphragm with light, observation of the trajectory in a cloud-chamber,[26] observation of movable slits placed immediately to the left of the diaphragm,[27] measuring the momentum of the diaphragm,[28] or, as Bohr himself suggested, measuring the momentum of a single-slit diaphragm placed to the left of the two-slit diaphragm.[29]

Bohr argues that if the single-slit diaphragm D_1 and the double-slit diaphragm D_2 were fixed rigidly to the apparatus and laboratory frame F, there could be no control of the momentum transferred between the object and the diaphragm, since the momentum would be irretrievably 'buried' in the apparatus (see Figure 5). If D_1 were not rigidly attached to the apparatus, it would be possible to measure its momentum before and after the passage of the object. This could be done in various ways, for example, by suspending D_1 from a weak spring and measuring its vertical deflection by means of a pointer attached to the apparatus support, or by applying the law of conservation of momentum to a collision between D_1 and some test body whose momentum was suitably controlled before and after the collision with D_1 (see Figure 6). But, in the former case, any reading of the scale, 'in whatever way performed', would involve an uncontrollable change in the momentum of the diaphragm which would render its position uncertain.[30] The same is true in the latter case: even if the position of D_1 were accurately

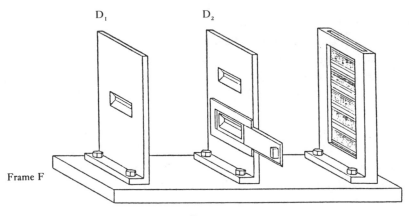

Figure 5
(By courtesy of Prof. P.E. Schilpp.)

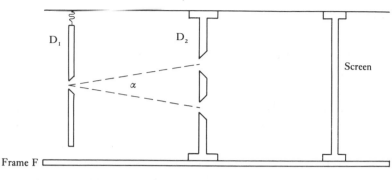

Figure 6

known before and after the first and second measurement of its momentum, its position during the interval between the two measurements would be uncertain, owing to its displacement by the test body.[31] The momentum transferred to D_I would be different, depending on whether the object passed through the upper or lower slit of D_2; if 'a' were the angle between the paths of the objects passing through the upper and lower slit, then this difference would be hka ('k' signifying the wave-number). Any control of the momentum of D_I which was sufficiently accurate to enable this difference to be determined would involve a latitude in the position of the diaphragm of the order of $1/ka$. Moreover, an uncertainty of this order in the position of D_I would involve an equal uncertainty in the positions of the fringes on the screen, since the number of fringes per unit length would be equal to ka; consequently no interference effect would be observable. The recoil of the diaphragm D_I, as the object passed through it, would have the effect of destroying the interference pattern. Apropos the analysis of this experiment Bohr writes:

> This point is of great logical consequence, since it is only the circumstance that we are presented with a choice of *either* tracing the path of a particle *or* observing interference effects, which allows us to escape from the paradoxical necessity of concluding that the behaviour of an electron or a photon should depend on the presence of a slit in the diaphragm through which it could be proved not to pass. We have here to do with a typical example of how the complementary phenomena appear under mutually exclusive experimental arrangements (cf. p.40) and are just faced with the impossibility, in the analysis of quantum effects, of drawing any sharp separation between an independent behaviour of atomic objects and their interaction with the measuring

instruments which serve to define the conditions under which the phenomena occur.[32]

We may either trace the trajectory of the objects, employing the particle model, or observe the interference effect, foregoing knowledge of the trajectories of the objects; but we may not do both.

What is the nature of Bohrian wholeness? It seems to be a sort of organic wholeness: the object and the instrument in mutual interaction form an organic whole in that the parts of the whole are intimately bound up together in causal interrelations. Bohr was well aware of the organic character of quantum-mechanical wholeness; indeed, it was this feature which led him to draw an analogy between a quantum-mechanical system and a living organism. The analogy, unfortunately, throws little light on the nature of Bohrian wholeness.

5.5
The nature of observation

In the late twenties Bohr wrote that it is a presupposition of our ordinary concept of observation that there is a relatively sharp distinction between the observed object and the observing subject (including any instrument of observation), and that the process of observation does not appreciably affect the observed object.[33] Perception, as ordinarily understood, is primarily a passivity, not an activity; it is the reception of some datum which, *qua* datum, is not itself affected by the process of reception. Owing to the quantum of action, however, observation involves an ineluctable disturbance of the object and precludes the possibility of any sharp distinction between the object and the instrument. Thus the classical conception of observation, implying as it does both a distinct separation between object and instrument and an absence of appreciable disturbance, is an idealisation: in atomic physics 'the usual idealization of a sharp distinction between phenomena and their observation breaks down':[34]

> . . . the finite magnitude of the quantum of action prevents altogether a sharp distinction being made between a phenomenon and the agency by which it is observed, a distinction which underlies the customary concept of observation . . .[35]

The sharp distinction which Bohr is referring to here is what I have called a 'dynamic distinction'.

Moreover, our 'forms of perception' fail in microphysics not just because the objects are not directly perceivable by means of the senses but, more importantly, because there is no clear-cut distinction between the object and the instrument of observation. A definite distinction between subject and object is, Bohr maintains, a precondition of the possibility of observation and of knowledge of an objective world: when that condition fails to hold, our ordinary forms of perception, which are also conditions of perceptual experience of an objective physical world, cease to be applicable, and the words 'observation' and 'object' lose their ordinary (classical) sense. He writes:

> . . . a close connection exists between the failure of our forms of perception, which is founded on the impossibility of a strict separation of phenomena and means of observation, and the general limits of man's capacity to create concepts, which have their roots in our differentiation between subject and object.[36]

This is obscure. I understand Bohr to mean that, where the boundary between subject and object becomes blurred, not only do our forms of perception (which are preconditions of perceptual experience of an objective world) fail, but also the conceptual scheme, upon which our conception of an objective world is based, breaks down: that boundary constitutes the limit of our powers of forming well-defined concepts of *objects*. Bohr is making a Kantian point: the notion of an objective order of experience presupposes the notion of a subjective order, an ability on the part of a subject of awareness to distinguish between a purely subjective and an objective order of experience. Where no distinction between the subject of awareness and the object of awareness is possible, no concept of an *object* is possible. Only if continuity obtains, in the sense that the content of subjective experience is connected and organised in a continuous manner, is it possible to form a concept of the content as an object. The requisite continuity is provided by the 'forms of perception' – the network of causal relations within the continuous manifolds of space and time. Similarly, where no clear distinction obtains between the objects of investigation and the instruments of observation, no well-defined conception of the object as an object of possible perceptual experience is forthcoming. He writes:

> The description of ordinary experience presupposes the unrestricted divisibility of the course of the phenomena in space and time and the linking of all steps in an unbroken chain in terms of cause and effect.[37]

In reality, however, no such description is possible. The classical notion of observation, then, is an idealisation. Observation always brings about a disturbance in the state of the object; hence if observation occurs, there can be no 'unambiguous definition of the state of the system'. Hence:

> The very nature of the quantum theory thus forces us to regard the space-time
> co-ordination and the claim of causality, the union of which characterizes the
> classical theories, as complementary but exclusive features of the description,
> symbolizing the idealization of observation and definition respectively.[38]

Since the classical notion of observation is an idealisation, physics itself is an idealisation. The classical theories can be unambiguously applied only in the limit where all actions involved are large compared with the quantum of action.[39]

Bohr does not hold, however, that there can be no observation of microphysical objects. Observation, albeit highly theory-laden, is possible, but only on the condition that a definite distinction between object and instrument can be made. Now although no sharp *dynamic* distinction between object and instrument can be made, in virtue of the indeterminable interaction, a *theoretical* distinction, a 'cut' can be made: the 'cut' consists in treating the object theoretically as if it were an object dynamically distinct from the instrument. Where to make the 'cut' is to some extent arbitrary, as Bohr pointed out from the outset: the cut may be made either between the original object and instrument, or between the composite object consisting of the original object *cum* instrument and some further instrument, and so on *ad infinitum*.[40]

Bohr compares the difficulty of securing objective observation in microphysics with a similar difficulty in psychology. The fact that the perceiving subject may be part of the content of his own consciousness limits the possibility of objective self-awareness. The very possibility of self-awareness entails a movable, or relative rather than absolute, boundary between subject and object. The arbitrariness of the cut between object and instrument is illustrated by the experience of orienting oneself with a stick in a dark room: when the stick is held firmly, it is no longer felt as a distinct object, but as an extension of the hand – touch seems to be located at the tip of the stick.[41] Moreover, just as awareness presupposes a subject which is not itself an object of awareness, so observation in physics presupposes an (ultimate) instrument which is not itself an object of investigation: some ultimate measuring instrument must be kept outside the system that is treated in quantum-mechanical terms.[42]

Since a distinction, however arbitrary, between object and instrument can be made, the quantum of action does not so much preclude observation as restrict the possibilities of observation. Bohr came to see that the distinction between object and instrument (the object-subject distinction) presupposed by the concept of observation could be made in quantum physics after all. Observation in quantum physics is, as it were, a generalisation of observation in the classical sense, in that the quantum of action imposes a restriction on the scope of observation, and also in the sense that, of the two classical conditions of observation – a sharp distinction between object and instrument, and no disturbance of the former by the latter – only one is satisfied, namely the sharp theoretical distinction:

> Strictly speaking, the idea of observation belongs to the causal space-time way of description. Due to the general character of the relation (2), however, this idea can be consistently utilized also in the quantum theory, if only the uncertainty expressed through this relation is taken into account.[43]

5.6
The 'cut' and the classical concepts

The fact that the instrument must be treated solely as an instrument if an observation is to be made entails that the measurement interaction is indeterminable, for if the interaction is to be determined, the instrument must be treated both as an object and as an instrument, which is impossible. It could be said, then, that observation and measurement are possible only on the condition that the measurement interaction cannot be determined. This point explains Bohr's obscure remark that the determination of the measurement interaction 'is excluded by the very nature of the concept of observation itself'.[44] Moreover, treating the instrument solely as an instrument is tantamount to describing it in classical terms; hence simply by treating the instrument in classical terms we are debarred from surveying the measurement interaction.[45]

Where to make the 'cut' is not a wholly arbitrary decision: it is subject to the condition that the classical treatment of the instrument be practically indistinguishable from a quantum-mechanical treatment; otherwise the 'cut' would involve a serious inconsistency. Treating the instrument in classical terms, though strictly incorrect, is legitimate provided the instrument is a sufficiently massive body, for in that case the effects which are due to the quantum of action will be imperceptible or negligible. It is, of course, the relation of correspondence that makes the classical treatment of the instrument to all intents and purposes consistent with a quantum-mechanical treatment. Thus we are free to make the 'cut' only 'within a region where the quantum-mechanical description of the process concerned is effectively equivalent with the classical description'.[46]

In his later writings Bohr continually stresses that whereas in classical physics there is no distinction in treatment between the object and the instrument, in that both are described in classical terms, in quantum physics this distinction is fundamental:

> . . . the observation problem of quantum physics in no way differs from the classical physical approach. The essentially new feature of the analysis of quantum phenomena is, however, the introduction of a *fundamental distinction between the measuring apparatus and the objects under investigation*. This is a direct consequence of accounting for the functions of the measuring instruments in purely classical terms, excluding in principle any regard to the quantum of action.[47]

This point is perfectly compatible with the point made earlier, viz. that in general no precise distinction in dynamical terms can be made between the object and the measuring instrument during the interaction between them.

Bohr may also believe that treatment of the instrument in classical terms is necessary not only if observation is to be objective, or even possible at all, but also if it is to yield a *definite*, unambiguous result. If the instrument is treated in quantum-mechanical terms, then after the interaction it is generally described by a state vector which is a superposition of states pertaining to the measured observable. The result of the measurement is definite only if the instrument is treated in classical physical terms; and of course the instrument can legitimately be treated in classical terms simply because it is a comparatively massive object, and hence the difference between the classical treatment and the quantum-mechanical treatment is vanishingly small. This is what Bohr has in mind when he says that 'only with the help of classical ideas is it possible to ascribe an unambiguous meaning to the results of observation'.[48] Definiteness of the results of measurement is secured through the use of macroscopic instruments such as rigid scales, robust clocks, and irreversible amplification effects:

> . . . all unambiguous information concerning atomic objects is derived from the permanent marks – such as a spot on a photographic plate, caused by the impact of an electron – left on the bodies which define the experimental conditions . . . the irreversible amplification effects on which the recording of the presence of

atomic objects rests rather remind us of the essential irreversibility inherent in the very concept of observation. The description of atomic phenomena has in these respects a perfectly objective character, in the sense that no explicit reference is made to any individual observer and that therefore, with proper regard to relativistic exigencies, no ambiguity is involved in the communication of information.[49]

Bohr's frequently-made point about the irreversibility inherent in the concept of observation is perhaps a reference to Kant's distinction between the irreversibility of the objective time order of sensory experience as against the reversibility of the subjective time order. Bohr is also alluding to the idea that observation involves an irreversible causal relation between an object and a subject in which the object affects the subject, while the subject has no appreciable effect on the object.[50] He came to see that this notion of observation has a place in quantum physics, since the disturbance of the object is a *consequence* of observation rather than an integral part of the process of observation itself.

It is perfectly clear that Bohr insists that the instrument must be described in classical terms, but it is not entirely clear what he means by 'classical terms'. This point needs further discussion.

5.7

The necessity of describing the instrument in classical terms

Bohr continually stresses that the instrument of observation must be described in classical terms:

> ... it is decisive to recognise that, *however far the phenomena transcend the scope of classical physical explanation, the account of all evidence must be expressed in classical terms*. The argument is simply that by the word "experiment" we refer to a situation where we can tell others what we have done and what we have learned and that, therefore, the account of the experimental arrangement and of the results of the observations must be expressed in unambiguous language with suitable application of the terminology of classical physics.[51]

What exactly does Bohr mean by 'classical concepts'? He may have in mind higher-level concepts such as 'magnetic field' or lower-level concepts such as 'position', 'velocity', etc., or both. He seems to make no distinction between what might be called 'theoretical' concepts and 'observational' concepts, or, if one prefers, between 'more theoretical' and 'less theoretical' concepts. He regards theoretical concepts as evolving as it were from the ordinary-language concepts which we use to describe our everyday perceptual experience: the notions of position, velocity, mass etc., as defined within classical physics, are simply more refined versions of everyday concepts. In his later writings he explicitly includes the physical concepts of ordinary language under the heading of 'classical concepts': he speaks frequently of the indispensability of 'everyday concepts, perhaps refined by the terminology of classical physics'.[52] Thus he writes:

> Notwithstanding refinements of terminology due to accumulation of experimental evidence and developments of theoretical conceptions, all account of physical experience is of course, ultimately based on common language, adapted to orientation in our surroundings and to tracing relationships between cause and effect.[53]

Bohr's point is that it is a necessary condition of *unambiguous* and *objective* communication that the experimental apparatus be describable in common-sense ordinary-language terms:

> . . . even when the phenomena transcend the scope of physical theories, the account of the experimental arrangement and the recording of observations must be given in plain language, suitably supplemented by technical physical terminology. This is a clear logical demand, since the very word "experiment" refers to a situation where we can tell others what we have done and what we have learned.[54]

In the late twenties Bohr states that 'all experience must ultimately be interpreted in terms of the classical concepts'.[55] Classical concepts are necessary for 'reducing our sense impressions to order', and for 'relating the symbolism of the quantum theory to the results of observation'.[56]

Bohr states that the correspondence principle was the outcome of his desire to utilise the classical concepts as far as possible in the quantum theory.[57] This view, however, is somewhat anachronistic: the earliest statement of the thesis of the indispensability of the classical concepts occurs in 1922. He writes:

> From the present point of view of physics . . . every description of natural processes must be based on ideas which have been introduced and defined by the classical theory.[58]

Bohr's early theory of atomic structure may have been motivated to some extent by a tacit desire to use the classical concepts as far as possible; but this motivation did not become explicit until later. By the late twenties he thought that it was misconceived to think that the problems of atomic theory could be obviated by replacing the concepts of classical physics with new concepts. Schrödinger was of the opposite opinion: in a letter of 5 May 1928 to Bohr he says that the fact that the description of the position and momentum of an object is possible only with a limited degree of exactness . . .

> . . . seems to me very interesting as a limitation of the applicability of the *old* concepts of experience. But it seems to me imperative to demand the introduction of *new* concepts, in which this limitation *no longer* occurs. Since what is unobservable in principle should not at all be contained in our conceptual scheme, it should not be representable in terms of the latter. In the *adequate* conceptual scheme it ought no more to seem that our possibilities of experience are restricted through unfavourable circumstances. But it would be very difficult to discover this new conceptual scheme, since as you have so impressively stressed, the necessary re-organisation involves the most profound levels of our knowledge, space, time and causality.[59]

Bohr replies:

> . . . I am scarcely in complete agreement with your stress on the necessity of developing 'new' concepts. Not only, as far as I can see, have we up to now no cues for such a re-arrangement, but the 'old' experiential concepts seem to me to be inseparably connected with the foundation of man's powers of visualising.[60]

Schrödinger sent Bohr's letter to Einstein, who remarked:

> The Heisenberg–Bohr tranquilizing philosophy – or religion? – is so delicately contrived that, for the time being, it provides a gentle pillow for the true believer from which he cannot very easily be aroused. So let him lie there.[61]

Einstein saw Bohr's position as conceptual conservativism.

Bohr's reply to Schrödinger is very revealing: our ordinary empirical concepts seem to him to be 'inseparably connected with the foundation of man's powers of visualising'. This squares with other statements from the same period. Thus, he says that it lies in the nature of observation that all experience must ultimately be expressed in terms of classical concepts, neglecting the quantum of action.[62] Our ordinary forms of perception are idealisations; nevertheless they 'colour our whole language'.[63]

There is an unmistakably Kantian overtone to all this: the classical concepts, which include the 'forms of perception', are indispensable for intersubjective communication about our perceptual experience. In the late forties he states that 'no content can be grasped without a formal frame' and that 'all knowledge presents itself within a conceptual framework adapted to account for previous experience'.[64] By 'a conceptual framework' he means a scheme which provides an 'unambiguous logical representation of relations between experiences'.[65] We have here the Kantian idea that while the content of our experience is given through the senses, its form is contributed by thought. Moreover, the formal elements of experience are the Kantian ones – spatio-temporal continuity and causal connection:

> The description of ordinary experience presupposes the unrestricted divisibility of the course of the phenomena in space and time and the linking of all steps in an unbroken chain in terms of cause and effect.[66]

The point seems to be that quantum mechanics can be applied and tested only through sensory experience, through observations on macroscopic instruments; and such instruments, *qua* macroscopic objects, must be describable in terms of the classical framework, i.e. as having locations within a spatio-temporal manifold, and as acting upon, and reacting against, each other.

It is doubtful, however, whether Bohr holds that the classical conceptual framework is *a priori* in the Kantian sense of expressing the necessary conditions of the possibility of human sensory experience; it is indispensable, rather, in the neo-Kantian sense that it is a framework which constitutes sufficient conditions for the possibility of objective experience. It may not be the only possible conceptual structure, but it is indispensable in the sense that we cannot jettison it lock, stock, and barrel; to do so would be to jettison experience as we have it. The classical conceptual scheme, however, can be modified, and indeed must be modified if we are to *extend* our experience; it must be *widened* to accommodate such extension:

> . . . all knowledge presents itself within a conceptual framework adapted to account for previous experience and . . . any such frame may prove too narrow to comprehend new experiences.[67]

We are reminded here of Otto Neurath's metaphor of the ship: while afloat, the ship cannot be rebuilt from the keel up; at most it can be modified plank by plank. Bohr's methodological precept here seems to be that in order to widen the classical conceptual framework, the modifications which we make to it should be as weak as possible.

It is worth pointing out here that, so far as I am aware, Bohr nowhere states unequivocally that microphysical objects, as distinct from macrophysical ones, must be conceived of in classical terms. He may have held, however, that it is very difficult to conceive of them in any other terms – a plausible view, for if we are to attain any intuitive understanding or conception of what microphysical objects are *like*, then we must conceive of them as being like objects with which we are already acquainted. By 'objects' I mean here entities which can be brought under concepts and such that we can imagine what it would be like to perceive them. Bohr's interest in retaining the classical concepts as far as possible in quantum physics stems partly from his taking seriously the hypothesis that there are microphysical *objects*; it stems also from his belief that modifications to physical theory should be made only to the extent that they are needed in order to account for new phenomena.

Paul Feyerabend has criticised the thesis that our perceptual experience must be described in terms of the conceptual scheme of classical physics. Our experience, he stresses, can be comprehended under quite different conceptual schemes, such as those of mythology, Aristotelian physics, etc.[68] Why not abandon the inadequate classical concepts and create new ones? Bohr's reply would be that theoretical progress in physics comes about through the continuous refinement of the concepts which have worked well up to now, and not by the creation of completely new concepts. Feyerabend, moreover, is assuming that by the term 'classical' Bohr has in mind solely the sophisticated concepts of classical physics; he is assuming also that by the phrase 'must be described', Bohr has in mind an *a priori* necessity in the Kantian sense. Both of these assumptions are mistaken: by 'must' Bohr has in mind the neo-Kantian notion of 'indispensable at present'; and he does not rule out further refinements of ordinary common-sense notions such as position, speed, etc.

Further consideration must now be given to the question of whether Bohr holds, as I have maintained he does, that microphysical objects are observable and measurable.

5.8
The microphenomenalist reading

Mario Bunge and others have interpreted some of Bohr's statements as indicating that he held that we do not observe microphysical objects at all: all we observe are macroscopic effects in macrophysical objects; indeed, there are *no* microphysical objects.[69] This interpretation goes by the name of 'microphenomenalism', since it treats microphysical objects in the same way as classical phenomenalism (macrophenomenalism) treats microphysical objects, viz. as non-entities. This reading is by no means utterly implausible. Bohr constantly stresses the importance of the macroscopic character of the instruments of observation, and the importance of their being describable in classical terms. Moreover, he states that when the word 'phenomenon' is used to refer exclusively to 'observations obtained under specified circumstances . . . the observational problem in atomic physics is free of any special intricacy, since in actual experiments all evidence pertains to observations obtained under reproducible conditions'.[70] The 'specified circumstances' include 'an account of the whole experimental arrangement'.[71] This sort of statement might be taken to mean that observation in quantum physics is no different from that in classical physics, since in both cases what we observe are exclusively macroscopic objects, there being no microscopic objects there to observe. But all that Bohr is intending to say is that a quantum-mechanical description of a microphysical object is well-defined only when the object is placed in the context of a well-defined experimental arrangement; indeed, a phenomenon in Bohr's sense might well be said to be the object as it appears in such a context.

While the microphenomenalist reading cannot be summarily dismissed, it is, I am convinced, mistaken. It is incompatible with Bohr's numerous analyses of experiments in which microphysical objects are taken to play an essential causal role: in these experiments the microphysical objects are treated as the *causes* of the macroscopic effects which we observe by means of the senses (e.g. electrons causing tracks in cloud-chambers, etc.).[72] Besides, wave-particle duality was a problem for Bohr simply because he did not adopt a microphenomenalist attitude to the classical models. The microphenomenalist reading rides roughshod over the entire subtle argument expounded in this chapter, making a mockery of Bohr's dogged struggle to find a way through the tortuous labyrinth of the philosophy of quantum physics.

5.9
Observation and objectivity

In his later writings Bohr stresses again and again that observation in quantum physics is perfectly objective. In some of these statements, however, 'objective' has the sense of 'intersubjectively valid' as distinct from 'revealing a pre-existing property of the object'. For example:

> The description of atomic phenomena has in these respects a perfectly objective character, in the sense that no explicit reference is made to any individual observer and that therefore, with proper regard to relativistic exigencies, no ambiguity is involved in the communication of information.[73]

I do not believe, however, that Bohr intends merely to say that observation in quantum physics is intersubjectively valid; he means also to say that it reveals properties of the object which are not created by the process of observation. What is the textual evidence for this assertion?

Bohr frequently cautions against talk of 'creation of physical attributes to objects by measurements', on the grounds that such a locution does violence to the ordinary meanings of the terms 'attributes', 'objects', 'observation', and is liable to be confusing.[74] The point is that if the process of observation creates the property that is observed, then talk of 'observation' here is a misnomer. Indeed, it is odd to say that we *observe* or *measure* the position, say, of an object, if the observed position is created by the act of observation; it is odd to speak of an *object* in this context, i.e. an entity whose properties are independent of the observing subject. Again, Bohr deprecates talk of 'disturbance of the phenomena by observation'(ib.) for the same reason: what we observe, in the ordinary sense of this term, is not a disturbed property. This is not to say that observation in quantum physics does not disturb the observed object; it does, but what we actually observe is the property of the object before it is disturbed, and not the property that results from the disturbance. (I shall take up this point again in Chapter 7.)

Bohr stresses these points, I believe, because he may himself have been confused by such locutions during the late twenties. At that time he may have been inclined to think that observed properties are subjective in the sense that they are somehow created by the process of observation. In 1929 he states that whereas in the theory of relativity the objective reality of observable objects is still rigidly maintained, in the quantum theory 'the classical ideal cannot be attained in the description of atomic phenomena'.[75] The 'classical ideal' is the 'ideal of objectivity', and 'objective' here has the sense of 'having to do with the inherent properties of the object'. He speaks also of the new light which the quantum of action throws on 'the old philosophical problem of the objective existence of phenomena independently of our observations':

> The limit, which nature herself has thus imposed upon us, of the possibility of speaking about phenomena as existing objectively finds its expression, as far as we can judge, just in the formulation of quantum mechanics.[76]

Statements such as these lend some support to the view that, at this time at any rate, Bohr held the observed properties of microphysical objects to be subjective in the sense that they are artefacts of the process of observation. After his painstaking analyses of the process of observation in numerous imaginary experiments, however, he came to see that there were no theoretical grounds whatever for the 'creation-by-measurement' view.

Thus, later on, he says that the notion of complementarity in no way involves 'a departure from our position as detached observers of nature'; rather, it expresses our situation 'as regards objective description in this field of experience'.[77] Commenting on this remark in a letter to Bohr, Pauli objected that it did not square with Bohr's earlier insistence that we are both actors and spectators, i.e. that we are not wholly detached observers (we affect what we observe). Bohr's reply is so important that it needs to be quoted at some length:

> A phrase like 'detached observer' . . . [used] in connection with the phrase 'objective description' . . . had to me a very definite meaning. In all unambiguous account [*sic*] it is indeed a primary demand that the separation between the observing subject and the objective content of communication is clearly defined and agreed upon . . .[78]

And in a further letter he states:

> To characterize scientific pursuit I did not know any better word than detachment . . . I wanted to stress the difficulties which . . . have had to be overcome to reach the detachment required for objective description or rather for the recognition that . . . we meet with no special observational problem beyond the situations of practical life to cope with which the word 'observer' has been originally introduced . . . The point which I especially wanted to stress . . . is that, just by avoiding any such reference to a subjective interference which would call for misleading comparison with classical approach, we have within a large scope fulfilled all requirements of an objective description of experience obtainable under specified experimental conditions.[79]

Again, in his 'Discussion with Einstein', he states that apropos the occurrence of individual events we ought not to say that we have to do with 'a choice on the part of the "observer" constructing the measuring instruments and reading their recording', for it is not possible for the observer 'to influence the events which may appear under the conditions he has arranged'.[80]

Admittedly, the textual evidence which I have cited in support of my reading is not so unambiguous as to confirm the reading beyond the shadow of a doubt. Indeed, the evidence can be construed in a way that agrees with the microphenomenalist reading: Bohr's point may be that we do not create what we observe, or disturb it by observing it, simply because what we observe are exclusively massive, macroscopic objects. My reading, however, squares with *all* the textual evidence: in particular it squares with the theory of observation and measurement that is presupposed by Bohr's explanation of the uncertainty principle (the theory of kinematic-dynamic complementarity). How *could* Bohr have given that explanation if he was a microphenomenalist? Moreover, my reading credits Bohr with a comprehensible theory of observation and measurement. Indeed, I know of no plausible account, either qualitative or formal, of measurement in terms of the 'creation-by-measurement' view.

Bohr, then, held what I shall call the *objective-values theory* of measurement, according to which successful observation or measurement reveals the objective, pre-existing value of an observable. This theory accords with Bohr's explicit statements about measurement. For example:

> . . . we must recognise that a measurement can mean nothing else than the unambiguous comparison of some property of the object under investigation with a corresponding property of another system, serving as a measuring instrument, and for which this property is directly determinable according to its definition in everyday language or in the terminology of classical physics.[81]

This comparison can be made only if there is a 'coincidence in space and time of the object and the means of observation', a coincidence which involves an interaction between the object and the instrument.[82] Thus when Bohr states that 'the observation problem of quantum physics in no way differs from the classical physical approach',[83] what he has in mind is the objective-values theory.

My contention that Bohr held the objective-values theory is controversial: it will be denied by those hostile critics of Bohr (such as Popper) who allege that his theory of measurement is subjectivist, and also by some of those who are sympathetic to Bohr's views. It is important here to note that the objective-values theory should not be confused with what I shall call the *intrinsic-values theory* of properties (or the 'intrinsic-properties theory'), according to which all the observables of an object have, at any moment, definite values. The latter theory, which Bohr rejected, is logically independent of the former. Confusion between these two distinct theories may have misled some into thinking that Bohr did not hold the objective-values theory. I shall examine the objective-values theory in the next chapter, and shall discuss the intrinsic-properties theory in Chapter 7.

5.10
A brief assessment of Bohr's argument

Is Bohr's argument for kinematic-dynamic complementarity cogent? It is certainly very persuasive. Its cogency, however, depends upon the truth of the two main premisses: (*a*) the exact measurement of canonically conjugate observables requires mutually exclusive experimental arrangements; and (*b*) the interaction between the object and the instrument of measurement is indeterminable. Premiss (*a*) is a universal generalisation which, by its very nature, is difficult to establish demonstratively. Nevertheless, Bohr's argument for it – based as it is on his thesis that a position measurement requires an instrument that is fixed with respect to the spatial reference frame, whereas a momentum measurement requires an instrument which is loose with respect to that same frame – is very powerful. I know of no convincing counter-example to it. Given premiss (*a*), Bohr's argument for premiss (*b*) is also very persuasive; even Einstein conceded defeat in his tenacious struggle to refute it. Bohr's argument, based on these two premisses, not only provides a plausible justification of the uncertainty principle, but also explains exactly and in detail how the uncertainty, and hence kinematic-dynamic complementarity, arises. I know of no other argument that succeeds in doing this as well as Bohr's: his argument is a *tour de force*, and it constitutes his most original, and most important, contribution to the interpretation of quantum mechanics. While going to the very heart of his interpretation, the argument does not comprise all that he has to say concerning kinematic-dynamic complementarity: there remains the question of the ontic or semantic construal of that notion; but I shall go into that question in detail in Chapter 7.

6
Bohr's
theory of measurement

I propose now to examine Bohr's informal theory of measurement from a somewhat more formal point of view, and to consider how his theory relates to the orthodox theory of measurement in quantum mechanics. This is a difficult task, since Bohr nowhere discusses the theory of measurement in formal terms. My discussion, therefore, will be less closely tied to Bohr's own statements. I shall also consider some important objections to the theory.

6.1
The objective-values theory of measurement

Before we look into this theory, it will be useful to have a brief sketch of some of the relevant formal elements of quantum mechanics.

In the Hilbert space formulation of quantum mechanics the measurable physical properties ('observables') of an object are represented by linear Hermitian operators A, B, . . . which have complete orthonormal sets of eigenvalues a_n, b_n, . . . corresponding to the subspaces (closed linear manifolds) of a Hilbert space H. These operators are in one-to-one correspondence with the projection measures P^A, P^B, . . . on the Hilbert space such that $P^A = \Sigma_n a_n P_n$, where P_n are the projections of the eigenvectors ψ_n belonging to the eigenvalues a_n of A, and a_n are discrete, non-degenerate values (for simplicity I shall deal only with operators which have such values, and shall denote observables and the corresponding operators by the same symbols). The states are represented by vectors, or rays ψ, ϕ, . . . in H, or by linear operators W, W', . . . defined everywhere on H (the statistical or density operators). There is no universal agreement about what the state vector signifies. Some regard it as representing the real physical state of an individual object (the ontic interpretation); others take it to denote a state of knowledge of an object (the epistemic interpretation); and others still take it to signify the state of a statistical ensemble of objects (the statistical interpretation). I shall speak of the 'state of an object' without intending to prejudge the question of interpretation. Exactly how Bohr regarded the state vector is a question I shall address below.

The observables and states in quantum mechanics are related by probability measures. The probability that an observable A of an object in state ψ or W has, or will be found on measurement to have, the value a_k is: prob $(\psi; A, a_k) = |(\psi_k, \psi)|^2$, or prob $(W; A, a_k) = \mathrm{Tr}\ (WP^A_{a_k})$, where '$\psi_k$' is the eigenvector corresponding to the eigenvalue a_k, and '$P^A_{a_k}$' is the projection of A onto the subspace corresponding to a_k, and 'Tr' is the trace operation. The temporal evolution of the state of an object, and hence of the probabilities, is specified in a strictly deterministic manner by a partial differential equation, the Schrödinger equation. The theory is indeterministic in the sense that the probability functions which relate states to observables generally do not have the value one.

Measurement is a species of observation: it is the determination by observation of the magnitude of a quantity corresponding to some physical property of an object. Assuming that in general the measured value of an observable is the value that the observable possessed immediately prior to the measurement, how is the measurement process to be described?

The value of an observable of a microphysical object cannot be determined directly, by sensory observation; it can be determined only indirectly, by sensory observation of the value of some macroscopic observable of a macrophysical object, from which the value of the observable in question can be inferred. In order that this inference can be made, the possible values a_n of an observable A (say) of the microphysical object O must be in one-to-one correlation with the possible values g_n of an observable G of the macrophysical object M: it must be possible to bring O and M into a suitable physical relation such that when the relation obtains, the value of G is g_k if and only if the value of A is a_k. For example, G may be the position of a pointer on a scale. The physical relation is a *causal* interaction between O and M such that G is caused to take on the value g_k at time t' when A has the value a_k at t ($t < t'$). Bohr was well aware of this point: he states that 'it should not be forgotten that the concept of causality underlies the very interpretation of each result of experiment, and that even in the coordination of experience one can never, in the nature of things, have to do with well-defined breaks in the causal chain.'[1] This statement makes it clear that Bohr did not believe that quantum mechanics precludes strictly deterministic causal relations between objects; what it precludes in general are deterministic predictions of the sort that are available in classical mechanics, i.e. deterministic predictions of the state of an object at some future time, given our knowledge of its state at some previous time. The very idea of measurement in quantum mechanics presupposes that deterministic causal relations between the object and the instrument obtain. It is important to note that measurement presupposes strict causality, but not strict continuity, i.e. that every causal change is continuous in the sense that during the change the object passes through all of the states in the continuous series of states which occupies the interval during which the change takes place. There is, of course, no inconsistency in saying that the measurement process is causally determinate and also that quantum mechanics is in a certain sense an indeterministic theory, for it is logically possible that there should be determinate causal relations between two or more objects even though some changes in the physical states of a single object are not deterministic.

Since O and M interact, O inevitably disturbs M, and M disturbs O. The disturbance of M is necessary for the measurement. And while O also is disturbed, this disturbance does not enter into the knowledge of the value of the observable which the measurement process provides: the measured value is not a 'disturbed' value in the sense that it is produced by the measurement process itself; it is the value of the observable immediately before the state of the object is disturbed by the measurement interaction. If the value observed were produced by the interaction between the object and the instrument, then it would be quite inappropriate to speak of a *measurement* taking place.

Since the interaction between O and M involves an exchange of energy or momentum, it is describable in quantum-mechanical terms. The interaction is represented by a unitary transformation on the Hilbert tensor product space, $H = H^O \times H^M$, of the composite object (or 'system') O + M. Different sorts of transformations lead to different sorts of correlations between the values of A and G. Two sorts of transformation are compatible with the objective-values theory of measurement. The first sort of transformation correlates the initial (i.e. premeasurement) values of A with the final (i.e. post-measurement) values of G; it is defined by $V(\psi_i \phi_j) = \psi_j \phi_i$. Thus where the initial state of O is $\Sigma_n c_n \psi_n$, and the initial state of M is ϕ_k, the initial state of O + M is $(\Sigma_n c_n \psi_n)\phi_k$, and the transformation is:

$$V(\Psi) = \Psi' = \Sigma_n c_n V(\psi_n \phi_k) = \Sigma_n c_n \psi_k \phi_n = \psi_k (\Sigma_n c_n \phi_n).$$

In this case the final state of O + M is a product of the states ψ_k and $\Sigma_n c_n \phi_n$. This sort of transformation corresponds to what is generally known as an 'ideal measurement', since it renders the disturbance of the object O determinable: the initial value of A is transformed into the value corresponding to the initial value of G; hence the final, as well as the initial, value of A is determinable.[2]

The second sort of transformation is defined by $U(\psi_i\phi_j) = \psi_{j(i)}\phi_i$, where for each positive integer j, $j(x)$ is a function from the positive integers into themselves such that, if for some x, $j(x) = j'(x)$, then $j = j'$, and such that

$$\{\psi_{j(i)}\}_{i=1,2,\ldots} = \{\psi_n\}_{n=1,2,\ldots}.$$
$$j = 1, 2, \ldots$$

Thus if the initial state of O is $\Sigma_n c_n \psi_n$, with ψ_n a complete set of eigenfunctions of A, and the initial state of M is ϕ_k, then the initial state of O+M is $\Psi = (\Sigma_n c_n \psi_n)\phi_k$, and the final state of O+M is $\Psi' = U(\Psi) = \Sigma_n c_n \psi_{k(n)}\phi_n$. This transformation correlates the initial state of M, viz. ϕ_k, with the set of final states of O, viz. $\Sigma_n c_n \psi_{k(n)}$, and the initial state of O, viz. $\Sigma_n c_n \psi_n$, with the final state of M, viz. ϕ_n. Thus the initial value of G is correlated with a set of final values of A; and hence the final value of A is not known. More important, however, is the fact that the final value of G is correlated with the initial value of A.[3]

The objective-values theory, then, can be formulated in terms of the transformation theory of quantum mechanics. There is a problem, however,: the final state of O + M is not in general a product of a state of O and a state of M; and so it seems that from the point of view of O + M neither O nor M can be assigned a definite final state. The fact that the object is not in a definite *final* state is not necessarily a difficulty for Bohr's theory of measurement, since Bohr holds that the final state of the object is indeterminable; but the fact that the instrument is not in a definite final state is a genuine difficulty – indeed, it is the notorious 'measurement problem – for, if the instrument is not in a definite final state, then no definite result should be observable; yet, contrary to the theory, a definite result *is* observable. This difficulty is not peculiar to Bohr's theory of measurement, but a solution to it is required if Bohr's theory is to be wholly successful.

6.2

The measurement problem

The problem is that if the process of measurement is described by a unitary transformation, then the final state of the instrument is described by a linear sum, or superposition, of state vectors. In the examples of measurement in Section 6.1, the final state of the instrument, for transformations V and U, is the superposition $\Sigma_n c_n \phi_n$. The final state of the instrument, moreover, is a superposition even if its initial state is a definite known eigenstate of the instrument observable. If it is assumed that a superposition of state vectors does not represent a definite state, then it seems that the instrument cannot be said to be in a definite state once the measurement interaction has ceased, and consequently the pointer on the instrument cannot be said to be at rest at any definite position which displays a definite measurement result. The problem would not arise if there were suitable transformations which mapped initial pure states into final mixed states. There are, however, no such transformations. It seems, then, that according to quantum mechanics the measurement process should yield no definite outcome, or in other words, that measurement should be impossible. Yet, as we believe, successful measurements are actually made; so, contrary to the theory, measurement *is* possible.[4]

How would Bohr have dealt with this problem? He would, I suggest, have responded along the following lines. The problem arises because the measurement process is described in terms of a unitary transformation on the state of the compound object, $O + M$. But to treat the process in this way is to elide the distinction between the object and the instrument: it is to treat the instrument as an object, an object which is an integral part of a larger object. Treating the instrument as an object, however, precludes our treating the interaction between the object and the instrument as a process of *measurement*. The measurement problem thus arises as a consequence of our treating the instrument as an object. But even if we confined our attention to the object alone, the measurement process would not be fully describable in quantum-mechanical terms. We might treat the evolution of the state of the object alone by means of an operation describing a perturbation on the Hamiltonian operator associated with the energy of the object, but only if we had knowledge of the perturbation, which of course we do not have, and cannot have if a measurement is to be made. Thus it is not surprising that the process of measurement cannot be properly described in terms of a unitary transformation. If we treat the instrument as an object, and describe the object-instrument interaction in quantum-mechanical terms, then measurement *is* impossible.

Bohr's theory does not entail that the final state of the instrument is indefinite if the object-instrument action is described in quantum-mechanical terms; it does, however, make this result unsurprising. Bohr, as we have seen, insists that it is a condition of successful measurement that the interaction between the object and the instrument of measurement is not surveyable, and is not treated in quantum-mechanical terms. This point is borne out by the formalism of the theory: if the measurement interaction is described by a unitary transformation of the state vector describing the combined system, then after the transformation neither the object nor the instrument individually can in general be assigned a definite state vector, since the vector describing the combined system cannot in general be factorised into two distinct vectors belonging to the object and the instrument. According to quantum mechanics, once they have interacted, the object and the instrument form an indivisible whole. Bohr's theory, then, cuts the measurement problem off at its roots: it prohibits from the outset the sort of theoretical treatment of the process of measurement that leads to the problem. It is for this reason that nowhere in his writings does Bohr give a formal account of the process of measurement. The Bohrian solution to the measurement problem is brilliantly simple; but it gives rise to an acute puzzle of its own, for how is it that the pointer of a macroscopic instrument may display a definite position, say, when the instrument is described in classical physical terms (when we simply look at it), yet have no definite position when we treat it as a quantum-mechanical object (foregoing inspection of it). A satisfactory solution to this puzzle is required if Bohr's theory is to be successful.

6.3
The solution to the Bohrian measurement puzzle

The solution to the puzzle lies in Bohr's thesis that the ultimate instrument of measurement, i.e. the apparatus which the observer looks at or listens to, must be a comparatively massive macroscopic object which is describable in everyday physical terms. It is a necessary condition of a measurement's being made that the instrument should be comparatively massive; if it were not, then no result would be observable. But it is more than a necessary condition: the comparative massiveness of the instrument ensures that a definite result is observable. Why does it? Simply because the massiveness of the instrument guarantees that the 'interference' terms contained in the superposition state may for all practical purposes be ignored, since they have no observable effects: the superposition state of a sufficiently massive object is practically indistinguishable from a mixed state. Antonio Daneri, Angelo Loinger and Giovanni Prosperi have proposed a detailed explanation of how this comes about.

These authors argue that the final superposition state of the instrument is transformed into a state which is indistinguishable from a mixture; transformation, moreover, is due to a real physical process, an ergodic process that takes place solely within the instrument after the interaction with the object has ceased. They reject the projection postulate as John von Neumann understands it, since, they hold, the transformation is not an instantaneous and discontinuous process. On their view the measurement interaction triggers an amplification process which induces in the instrument a metastable state that evolves into a state of equilibrium, a definite state uniquely correlated with a definite state of the object.

They consider the state of the instrument to be quantised in a large number of different stationary states, which are divided into 'shells', each containing all the states within a given energy interval; the states in turn being subdivided into 'cells' corresponding to the observable values of the macroscopic observables in question. The measurement interaction transforms the instrument into a superposition of states corresponding to definite groups of cells. If certain ergodicity conditions are imposed, the superposition state evolves towards a state corresponding to a cell of maximum statistical weight, a state of equilibrium in which the phase correlations between the co-efficients of the superposition states practically disappear. This result is obtained only on the condition that the instrument is a sufficiently large macroscopic body. Given this condition, the result obtains even for instruments which do not assume thermodynamically metastable states. The state which results from the ergodic process is practically indistinguishable from a mixture, even though, strictly speaking, it is still a superposition.[5]

A similar result has been demonstrated by Wolfgang Weidlich. Given certain assumptions about the macroscopic structure of the instrument, Weidlich shows that the 'interference' terms in the superposition of states converge to zero after the interaction (in the sense of weak operator convergence). Although the superposition is not completely eliminated, since the Hilbert–Schmidt norm of the terms remains constant, it can be regarded as being a very close approximation to a mixture, the difference between the superposition and the mixture being indistinguishable in practice.[6]

These results lend support to Bohr's theory of measurement: they help to explain why it is that the macroscopic and classically describable character of the measuring instrument ensures the observability and definiteness of the results of the process of measurement. Bohr constantly stresses that measurement of microphysical observables is possible only by means of irreversible effects in macroscopic bodies:

> As all measurements thus concern bodies sufficiently heavy to permit the quantum to be neglected in their description, there is, strictly speaking, no new observational problem in atomic physics. The amplification of atomic effects, which makes it possible to base the account of measurable quantities and which gives the phenomena a peculiar closed character, only emphasises the irreversibility characteristic of the very concept of observation.[7]

The irreversibility of the amplification process gives the phenomenon '. . . that inherent feature of completion which is demanded for its well-defined interpretation within the framework of quantum mechanics.'[8] By the terms 'closed character' and 'feature of completion' Bohr is referring here to what he sees as necessary conditions of successful measurement: the observed property must be (*a*) definite, in the sense that it is not a superposition of properties; (*b*) determinate, in the sense that whether or not the object has the property in question at a certain time depends solely on the conditions obtaining at that time and the immediate past. Bohr holds that the definiteness and determinateness of observed properties are ensured by the irreversible amplification process.

The Daneri–Loinger–Prosperi theory has been criticised by Jeffrey Bub, and by J.M. Jauch, E.P. Wigner and M.M. Yanase. Bub objects that the Daneri–Loinger–Prosperi theory is actually contrary to Bohr's views, since it treats classical mechanics as an approximation to quantum mechanics: Bohr, however, holds that there are no quantum objects, and that quantum mechanics is concerned only with 'deviant' behaviour of classically describable macroscopic objects; and so it makes no sense to say that classical mechanics is an approximation to quantum mechanics.[9] This objection, however, is based upon a mistaken interpretation of Bohr's position. As we have seen, Bohr does maintain that there are quantum objects in the sense of microphysical objects which are correctly describable only in quantum-mechanical terms, and that classical mechanics is an approximation to quantum mechanics in the sense that the latter is a rational generalisation of the former. Jauch and his colleagues object that the Daneri–Loinger–Prosperi theory cannot explain negative result experiments.[10] This objection misses the mark, since such experiments do not constitute measurements of the sort that the theory is intended to explain, i.e. measurements which involve an interaction between the object and the measuring instrument.

Daneri, Loinger and Prosperi acknowledge the similarity between their theory and Bohr's interpretation of quantum mechanics.[11] Rosenfeld, supporting their theory, states that it is 'in complete harmony with Bohr's ideas'.[12] He writes: 'There should now remain no doubt that the reduction of the initial state of the atomic system has nothing to do with the interaction between this system and the measuring apparatus . . .'[13] In his last writings Rosenfeld expressed a preference for the theory of Ilya Prigogine, which is based upon a quantum-statistical theory of macrophysical systems rather than on ergodic theory.[14] In this theory the statistical operator is treated as a 'supervector' in a superspace defined as the direct product of the Hilbert space with itself, and the temporal evolution of the supervector is governed by the Liouville equation. When the state of the measuring instrument is described by such an operator, the phase relations between its components are eliminated, being projected into the subspace of the Hilbert superspace. Again, the final state of the instrument approximates asymptotically to a definite state, and hence any indefiniteness that may be present is not detectable in practice.[15]

The significance of Bohr's stress on the macroscopic, massive character of the instrument of measurement is now clearer from a formal point of view. If the instrument is a massive, macroscopic object, it can be described in classical terms in a way that allows the quantum-mechanical features of its description to be ignored.[16] Strictly speaking, of course, the 'interference' terms which are implicit in the state vector of the instrument never completely vanish, and measurement does not yield an absolutely definite result. But if measurement is to yield a quasi-definite, 'unambiguous' result, then any indefiniteness in the final state of the instrument must be practically undetectable. What the Daneri–Loinger–Prosperi theory and the Prigogine–Rosenfeld theory do is to explain in detail why the result of measurement appears to be definite when in fact it is not.

The fact that from the point of view of quantum mechanics an object does not have an absolutely exact position or an absolutely exact momentum, where 'absolutely exact' is defined in terms of the real number system, should not concern us, since the absolute exactness in question is ineluctably beyond the reach of any possible empirical observation. It is interesting to note, as Paul Teller has pointed out, that quantum-mechanical observables, unlike classical-mechanical observables, cannot in general be said to have absolutely exact theoretical values: continuous quantities, such as position and momentum, are represented by operators which have no eigenvectors with eigenvalues in the continuous part of their spectrum.[17] Hence there are no quantum-mechanical states in which continuous quantities have absolutely exact, point-values corresponding to non-rational real numbers; and so there are no states in which an exact value of such a quantity has the probability of 1. By contrast, classical mechanics contains descriptions of state in which continuous quantities have point-values. Although the inexactness that pertains to continuous quantities in quantum mechanics is not necessarily of the same sort as that which is involved in the superposition of states, it squares with the view that the classical conception of the absolute exactness of physical quantities is an idealisation.

6.4
Bohr's interpretation of the state vector

What, exactly, is Bohr's interpretation of the state vector? Teller, who treats this question in detail, argues that Bohr held a statistical interpretation of the state vector, to the effect that the vector describes the statistical distribution of an ensemble of phenomena, i.e. total, similarly prepared experimental arrangements described in terms of classical physics or ordinary language, the characteristics in question being macroscopic.[18] Teller notes that this conception of the ensembles conflicts with Bohr's realist construal of the notion of atoms, electrons etc., but he argues that the conflict is more apparent than real. Since I have already argued that Bohr does not hold that quantum mechanics refers only to macrophysical objects, I shall not press my disagreement with Teller on this point.

The question remains whether Bohr holds that the state vector describes ensembles of individual microphysical objects in well-defined experimental contexts. In support of the statistical reading, Teller cites statements such as the following:

> In the treatment of atomic problems, actual calculations are most conveniently carried out with the help of a Schrödinger state function, from which the statistical laws governing observations obtainable under specified conditions can be deduced . . .[19]

This statement, however, does not unequivocally convey the statistical interpretation. Bohr frequently states that quantum mechanics is a statistical theory.[20] But his point in such statements may be merely that quantum mechanics is not a deterministic theory: the term 'statistical' may, for Bohr as for many other physicists, have the sense of 'probabilistic', 'non-deterministic'. Bohr's point, then, may merely be that the state vector provides only probabilistic information about individual objects. At least one statement calls clearly for the statistical interpretation: in 1929 he states that

> . . . one might perhaps say that quantum mechanics is concerned with the statistical behaviour of a given number of atoms under well-defined external conditions . . .[21]

Note, however, the tentative way in which Bohr makes this point (the 'perhaps'). The textual evidence, then, does not unequivocally support the view that Bohr accepted the statistical interpretation of the state vector.

Bohr, I believe, was chary of the statistical interpretation, and that for two main reasons. First, he held that there is merely a conceptual difference between the statistical interpretation and the epistemic interpretation, but no substantive difference. His disagreement with Einstein concerned not so much the statistical interpretation as the question of completeness, the question of whether the state vector provides a complete description of an individual object. Contrary to Einstein, Bohr held that the information about an object which is provided by the state vector is complete, as complete as can be. Thus, while Bohr did not think that the statistical interpretation was mistaken, he saw no point in insisting upon it, since the state vector provides as complete a description of an individual object as is possible. Second, while Bohr did not think that the statistical interpretation of the state vector was mistaken, he believed that on its own it did not provide a complete basis for the interpretation of quantum mechanics: it did not do so since it did not provide an adequate formal basis for the derivation of the uncertainty principle. (My reading here is speculative, in the absence of suitable documentary evidence, but it is, I believe, correct.)

Some advocates of the statistical interpretation would agree with Bohr on this second point. Popper has argued since 1934 that if one adopts the statistical interpretation of the state vector, the uncertainty principle lacks a formal foundation in the theory. According to Popper the uncertainty relations should be interpreted as statistical scatter relations: they signify the relations between the scatter or dispersion of the values of observables in a statistical ensemble. Thus, if we select an ensemble of objects whose positions, say, lie within the range Δx, then the values of the momenta of these objects will be found to lie within the range Δp_x, and the product of these ranges will be $\Delta x \Delta p_x \geqslant \frac{1}{2} h$. These relations entail that if the position values are scattered over a comparatively narrow range, then the momentum values will be scattered over a comparatively broad range, and vice versa. For example, if we select only those particles which pass through a comparatively narrow slit in a diaphragm, and subsequently measure the momentum of each particle in this collection, then the momentum values will be scattered within the range Δp_y, and the narrower the width of the slit Δy, the broader the dispersion Δp_y. The wide dispersion of the momentum values, Popper holds, is due to the interaction between the objects and the diaphragm. Popper maintains, quite correctly, that, interpreted in this way, the uncertainty relations do not entail the uncertainty principle.[22] Popper's account of the statistical interpretation is very plausible: it makes clear how the relations between Δq and Δp arise, viz. through the disturbance of the objects brought about by the preparation process.

Those who hold Popper's view argue that, given the character of the formal derivation of the uncertainty relations, the statistical interpretation is the only possible, or at least the only plausible, interpretation of them. In the standard derivation which is due to H.P. Robertson, the uncertainty relations have the general form:

$$\Delta_W A \Delta_W B \geqslant \tfrac{1}{2}[\mathrm{Exp}_W(AB - BA)].$$

where 'Exp' signifies the expectation-value, or mean-value, functional. Where 'A' and 'B' are the position and momentum operators, the relations have the form $\Delta q \Delta p \geqslant h/4\pi$.[23] The expression '$\Delta_W A$' is taken to signify the standard deviation (or dispersion) of A in state W, or the root-mean-square deviation, where the mean-square deviation (or variance), $\mathrm{Exp}_W\{[A - \mathrm{Exp}_W(A)]^2\}$, represents the average deviation of A from its mean value in state W. Now it is natural to interpret the expression '$\Delta_W A$' as signifying the standard deviation or statistical dispersion of A, if the functional 'Exp' is read as the mean-value functional; and the latter reading is plausible, since 'Exp' has the same formal properties as the mean-value function in the classical theory of probability. But this reading is not mandatory, since the functional 'Exp' is derived by way of the quantum-mechanical probability function, and this in turn is defined in terms of the state vector. Thus the interpretation of 'Exp', and consequently of '$\Delta_W A$', depends upon one's interpretation of the state vector; and, as we have seen, the state vector can be interpreted in different ways. It is possible, therefore, to interpret the expression '$\Delta_W A$' either as the statistical dispersion of the values of A in an ensemble (the statistical interpretation of W), or as the uncertainty in our knowledge of the value of A in state W (the epistemic interpretation), or as the real indefiniteness of the value of A in state W (the ontic interpretation).[24] Popper's interpretation, then, while plausible, is not binding. If, moreover, one adopts the epistemic interpretation, then the orthodox interpretation of the uncertainty relations, viz. the uncertainty principle, follows naturally.

Besides, even if one adopts the statistical interpretation of the uncertainty relations, it does not follow from this that the uncertainty principle has no formal basis in the theory. Indeed, it is a consequence of the properties of the Fourier transform of an operator and of a state vector that there are no joint probability measures for the values of non-commuting observables in a single state, and non-commuting operators, such as q and p, can have no joint values in any single state. Thus,

$$P_{\psi}^q(E) \wedge P_{\psi}^p(F) = 0,$$

where 'P_{ψ}^q' is the projection of the operator q on the vector ψ, and 'E' is a bounded Borel subset of the set of real numbers, and '$\wedge$' is the greatest lower bound of the projections P_{ψ}^q and P_{ψ}^p. This equation is a consequence of the fact that if P_{ψ}^q has a value in E, then the support of the function ψ is compact; in which case the support of its Fourier transform is the entire set of real numbers. The significance of this equation is that there are no states of an object in which both the position observable and the momentum observable have, or will be found on measurement to have, a value within a definite Borel subset of the set of real numbers, and also that there are no joint probability measures for simultaneous values of such observables. If these observables were jointly measurable, then quantum mechanics would be deficient since it would lack probability measures for such measurements.[25] The statistical interpretation of the state vector, then, does not require the abandonment of the uncertainty principle or the idea of kinematic-dynamic complementarity. Bohr's circumspection (if he had any) concerning the statistical interpretation of the state vector was not entirely justified; indeed, the statistical interpretation is quite compatible with his general interpretation of quantum physics.

Bohr's position on the question of the interpretation of the state vector can be summed up as follows. He recognised, quite correctly, that the state vector can be interpreted either in statistical terms or in epistemic terms, and he believed that for practical purposes it was immaterial which interpretation one adopted. His preference, however, was for the epistemic interpretation, since he thought that this interpretation was less misleading concerning the question of completeness. There is no evidence to suggest that he adopted the ontic interpretation of the state vector; indeed, all the evidence is to the contrary.

It may be argued that the measurement problem arises only if the ontic interpretation of the state vector is adopted: if the state vector describes an ensemble, then an object belonging to an ensemble which is described by a superposition may be regarded as being in a definite, though, unknown, state. Such an argument, however, would not be persuasive, since it can plausibly be objected that the significance of the superposition is that members of an ensemble described by a superposition are not in any definite state. What needs to be shown, therefore, is that the statistical significance of the superposition is such that members of such an ensemble may be held to be in definite, though unknown, states.

Now that we have a clearer picture of Bohr's theory of measurement from a more formal point of view, it will be useful to consider how it compares with the orthodox theory of measurement, the theory that is implicit in most textbooks of quantum mechanics. In order to deal with this question, I shall examine von Neumann's theory of measurement, since it was the first mathematically rigorous theory and the basis of the orthodox theory.

6.5
Von Neumann's theory of measurement

Von Neumann argued that the quantum-mechanical state of an object may evolve in either of two radically different ways:

(a) in accordance with the Schrödinger equation,

$$\psi \to \psi' = \exp(-iHt/\hbar), \text{ or}$$
$$W \to W' = \exp(-iHt/\hbar)W(\exp(-iHt/\hbar));$$

(b) in accordance with the transition,

$$\psi \to \psi' = \Sigma_n(\psi_n, \psi)\psi_n, \text{ or } W \to W' = \Sigma_n(W_{\psi_n}, \psi_n)P_{\psi_n}.$$

The transition (*a*) describes the temporal evolution of the state of an object when the object is not interacting with another object; this transition represents a continuous, reversible process. The transition (*b*) describes a discontinuous, irreversible change in the state of the object which takes place when the object is measured. Transitions of this sort transform pure states into mixtures, and are indeterministic in the sense that they determine only the probability $|(\psi, \psi_n)|^2$ with which the result of the measurement will be ψ_n: observation of the measuring instrument reveals which of the ψ_n obtains. If the result is observed to be ψ_k, then ψ_k is taken to be the state of the object immediately after the measurement. The discontinuous transition applies also to the state of the system comprising the object and the instrument, so that if the initial state of the system is $(\Sigma_n c_n \psi_n)\phi_0$, then the discontinuous transition yields the mixture $\Sigma_n |c|^2 P_{\psi_n \phi_n}$. Von Neumann's postulation of (*b*)-type transitions (the 'projection postulate' as it is now called) thus solves the measurement problem.[26] I shall refer to this transition as the 'reduction of state'.

The projection postulate, as von Neumann formulates it, applies only to maximal observables, i.e. to observables the measurement of which fixes the value of a set of commensurable observables. This difficulty has been removed by a generalisation suggested by Gerhardt Luders which takes account of non-maximal measurement. According to the Lüders rule,

$$W \rightarrow W' = P_{\psi_k} W P_{\psi_k} / \mathrm{Tr}(P_{\psi_k} W P_{\psi_k}).[27]$$

The more general postulate, however, faces further difficulties. First, the postulate seems *ad hoc* in that it is introduced into quantum mechanics solely to account for the measurement process. But it is singularly odd that interactions in general can be described in terms of the Schrödinger equation, yet the measurement process can not. Why should the measurement process require special treatment? Why should there be, as von Neumann maintains,[28] two distinct sorts of processes? Besides, the postulate has little explanatory force, since it immediately invites the question as to how the discontinuous, instantaneous reduction of state comes about. Von Neumann has no clear answer to these questions. There is, I believe, no *sui generis* physical process corresponding to the projection postulate: the physical process involved in measurement is an ordinary interaction. Second, exactly when the reduction of state takes place affects the evolution of the state vector and the expectation values governed by it. Yet the theory does not specify the instant at which the reduction takes place: it might take place at any instant within the finite time-interval during which the measurement takes place, or at the instant when the measurement begins or ends; in each case the expectation value will not be the same.[29] This difficulty is all the more striking in the case of negative-result experiments and delayed-choice experiments.

Many textbooks attempt to alleviate the first difficulty by removing from the projection postulate any suggestion that it represents a physical process. The postulate is treated simply as providing an answer to the question 'If a measurement of an observable A yields the definite result a_k, what is the state of the object after the measurement?' The answer is 'ψ_k, where ψ_k is the eigenfunction belonging to the eigenvalue a_k'. Thus it is laid down as a postulate that if a measurement of an observable A yields the eigenvalue a_k, then the state of the object immediately after the measurement is described by the subspace of the Hilbert space which is spanned by the eigenvectors belonging to a_k. Thus construed, the projection postulate is more plausible.

There is, however, a serious difficulty with this construal of the postulate. Von Neumann makes it a condition of his theory of measurement that if a second measurement of an observable A is made immediately after the first measurement, then the same value of A will be obtained. Measurements which satisfy this condition are generally called 'ideal' or 'measurements of the first kind'.[30] The projection postulate entails that measurements governed by it are ideal, for, if the state of the object immediately after the first measurement of A is ψ_k, corresponding to the eigenvalue a_k, then the probability that a second measurement will yield the value a_k is prob $(a_k/\psi_k) = |(\psi_k, \psi_k)|^2 = 1$. The difficulty is that ideal measurements are the exception rather than the rule.

Ideal measurements are certainly obtainable. For example, if a photon is polarised in the x-direction, it is likely to pass through a filter oriented in the appropriate direction; and if it does so pass, it can be expected to pass through an identically oriented filter placed immediately behind the first one. Von Neumann's example of an ideal measurement is based on the Compton–Simon experiment. He argues that, assuming that the mechanical laws governing electron–photon collisions obtain, we may measure the position and central line of an electron–photon collision by measuring either the path of the recoil electron or the path of the scattered photon, or both, and the results will be the same. Moreover, one measurement may be made immediately after the other. This example is singularly inappropriate, since the two measurements in question are not made on the same object, but on different objects; it does not address the cases where the two measurements are made upon the same object.

There are two sorts of ideal measurements. First, if the object O before the measurement is described by an eigenfunction ψ_m of the observable A to be measured and the instrument M is in the eigenstate ϕ_0 of an observable G, then the composite object O + M is described by the state vector $\Psi = \psi_m\phi_0$ (ϕ_0 is the state of M corresponding, say, to the zero reading of some pointer). The measurement transition, then, is: $\psi_m\phi_0 \rightarrow \psi_m\phi_m = \Psi'$. In this case the result of the measurement a_m must be regarded as the value possessed by A both before and after the measurement.[31] Measurements of this sort are sometimes called 'filters'. Second, if the object is not described initially by an eigenfunction of the observable A to be measured, but by an eigenfunction of some other observable, then the measurement transition is: $\xi\phi_0 \rightarrow \psi_m\phi_m$. The observed value a_m may be regarded as the value possessed by A after the measurement, but not before. Measurements of this sort are generally known as 'state preparations'. It is only preparations of the first sort, filters, that are properly called 'measurements'.

Not all measurements, however, are ideal; indeed, most are non-ideal, that is, such that a measurement of the same observable made immediately after the first measurement yields a value different from that of the first. In this case, if the object is initially in an eigenstate ψ_i of the observable A, and the instrument is in an eigenstate of the instrument observable G, then the measurement transition is $\psi_i\phi_0 \rightarrow \xi\phi_k$. In general, ξ is not equal to ψ_k or even orthogonal to it; ξ, moreover, may not be an eigenfunction of any observable. The observed result, however, may be said to be the value of A immediately before the measurement.

These points, in essence, were made by Lev Landau and Rudolf Peierls in 1931. Following Bohr's views, they argued that measurements are generally not ideal, and that a measurement is ideal only if the operator of the measured observable commutes with the Hamiltonian operator governing the energy of the interaction between the object and the instrument. The only observable which is susceptible of an ideal measurement, they argued, is the position co-ordinate, the interaction energy being a function of the co-ordinate.[32]

Construed as providing an answer to the question, 'What is the state of the object immediately after a measurement has been made upon it?', the projection postulate cannot be accepted, since the answer, 'the eigenstate corresponding to the measured value', is not generally correct. On what, then, are we to base our predictions about the future behaviour of the object? Bohr does not say, but an answer that is compatible with the objective-values theory is 'upon our knowledge of the initial state of the object'. If the value of

A that is obtained by measurement is a_k, then the initial state of the object may be said to be ψ_k, the eigenstate corresponding to a_k. We base our predictions, then, on the same information as that specified by the projection postulate. The initial state ψ_k generally provides only probabilistic information about the future state of the object, since the process of measurement generally changes the state ψ_k into a different state which is not known. The practical consequences of using the projection postulate, for all its theoretical inadequacy, are thus innocuous. Nevertheless the postulate, and the theory of measurement on which it is based, must be rejected. Indeed, the projection postulate, understood as implying that the result of a measurement characterises the object *after* the measurement, is one of the main reasons why the objective-values theory has not been more widely advocated.

Bohr's theory of measurement escapes the problems that beset von Neumann's theory: Bohr's theory has no need of the projection postulate; nor does he ever suggest that it does. Moreover, Bohr emphasises that measurements are generally not ideal.

6.6
The subjective theory of measurement

Since the cut between the object and the instrument is movable, as von Neumann recognised, the projection postulate can be applied either to the object, or to the system comprising the object and the instrument, or to the larger system comprising the object, the instrument and some further instrument, and so on. If a measurement is to be made, then the projection postulate must be applied somewhere; but exactly where is arbitrary. It is theoretically possible to include the sense organs and brain of the observer within the object system. If this were done, the projection postulate would have to be applied to that system, and hence, as von Neumann noted, the cut between the object and the instrument (in this case the observing subject) would have to be made between the brain and consciousness of the observer.

This possibility suggested to Fritz London and Edmond Bauer that the reduction of state is not a straightforward physical process at all, but one which involves the consciousness of the observer. Since the cut is movable, they argued, the object together with the measuring instrument and the sensory organs and brain of the observer may be treated as a composite object and described in terms of the Schrödinger equation; in which case the cut must be placed between this composite object and the consciousness of the observer. The consciousness of the observer could not be included within the object system if an observation were to be made, since observation requires a subject of awareness as well as an object of awareness; if it were so included, the consciousness of the observer would be in a superposition of states, and hence not a definite state of awareness at all. If, however, the cut is made at the brain of the observer, then the reduction of state must be due to an act of awareness on the part of the observer. The observer's consciousness therefore constitutes the ultimate boundary between object and instrument, between object and subject. But if the reduction of state is due to the conscious awareness of the observer in this ultimate case, might it not also be due to the observer's awareness in the ordinary case? London and Bauer suggested that the reduction of state is in every instance due to the observer's act of awareness: the state of the object is reduced as soon as the observer looks at the instrument and notes the result of the measurement. The reduction of state, then, has nothing to do with the interaction between the object and the instrument; it is due to a conscious act of observation.[33]

This theory has bizarre consequences. How, one wonders, does the observer's act of awareness determine the value of the measured observable – by a sort of psychokinesis? Where the measured observable is microphysical, it cannot be observed in the visual sense; what is visually observed in this case is a macroscopic observable of the instrument with which it has interacted. Somehow the observer's looking at the macroscopic instrument causes the microphysical observable in question to take on a definite value. But how exactly? Does the act of observation affect the macroscopic observable, and that in turn (in a purely physical way?) affect the microphysical observable? On this theory there is no escaping the conclusion that consciousness can determine some properties of a macroscopic object.

This paradoxical consequence is vividly illustrated by Schrödinger's 'cat' paradox. Schrödinger considers a chamber containing a cat and a tiny quantity of some radioactive substance whose half-life is one hour. If an atom of the substance decays, it activates a Geiger-counter which in turn releases a mechanism which breaks a phial of cyanide, killing the cat. At the end of one hour the state vector for the system of atom plus mechanism plus cat is a superposition of vectors, one of which represents the cat alive, the other the cat dead.[34] According to the subjective theory, at the end of the hour the state of the cat is indefinite, which seems to imply that the cat is neither definitely alive nor definitely dead; it becomes one or the other only when the observer opens the door of the chamber and looks inside. This consequence is highly counter-intuitive. Common sense tells us that at the end of the hour the cat is definitely alive or dead depending solely on whether or not an atom of the radioactive substance has exploded; and the observer's inspection simply reveals which of the alternatives has occurred. Assuming that we can make sense of the notion that the cat is neither determinately alive or dead, how, one wonders, does the observer's merely looking affect the state of the cat? If the observer, on inspection, sees that the cat is dead, his merely looking, it seems, kills the cat. Directly, one wonders, or indirectly, by causing an atom of the radioactive substance to explode at the moment of inspection?

I shall not pursue these questions, for the theory presupposes the projection postulate, and raises more problems than it solves; it is, besides emphatically rejected by Bohr. He constantly stresses that measurement in quantum mechanics is as objective as measurement in classical physics.[35] He also repudiates the view that observation creates physical attributes – the 'creation of physical attributes' thesis. He makes this point again and again, stating that in quantum physics terms such as 'observation', 'attribute' have their ordinary language meanings, which do not imply that observation creates the observed property, or that the observed property is a 'disturbed' property.[36] The measurement process is complete when the instrument registers an observable result; whether or not the result is actually observed by an observer is irrelevant, so far as the measurement process is concerned. The result may be recorded either by the instrument, or by some other instrument, and may be observed long after the event; and the measurement result characterises the object immediately before the measurement is made.

6.7
Difficulties with the objective-values theory

I wish now to examine some objections to the objective-values theory. Consider an ensemble of structurally identical objects prepared in the same state $W = P_{\sum_n c_n \psi_n}$, where the ψ_n are eigenfunctions of an observable B belonging to those objects with eigenvalues b_n. For simplicity's sake, let B have only two eigenvalues b_1, b_2, and two eigenfunctions ψ_1, ψ_2; thus $\psi = c_1 \psi_1 + c_2 \psi_2$. If we measure B on each member of a large collection N of such objects, there will be a dispersion in the measured values: a certain fraction x/N will have the value b_1, another fraction y/N the value b_2 ($N = x/N + y/N$). It seems that the objective-values theory requires us to take the ensemble E, to which the collection N belongs, as a mixed ensemble which is the weighted sum of the pure proper subensembles E', E'' determined by the eigenvalues b_1, b_2; thus $E = |c_1|^2 E' + |c_2|^2 E''$. If we construe the ensemble E in this way, it seems that the state vector $\psi = c_1\psi_1 + c_2\psi_2$ should be taken to represent a mixture.

It is generally held, however, that this cannot be done, since ψ represents a pure state, and not a mixed state: the statistical operator is a projection operator $W = P_\psi$, and this is taken to represent the state of a pure ensemble, and not the state of a mixed ensemble, since the state vectors in terms of which the projection operator is defined are all identical (up to phase factors). But since the state vector ψ can generally be expressed as a linear sum or superposition of state vectors ψ_1, ψ_2 ($\psi = c_1\psi_1 + c_2\psi_2$), ψ can surely be regarded as representing a state which is a mixture of states ψ_1 and ψ_2. The stock objection to this suggestion is that a superposition cannot represent a mixed state. The grounds for this thesis are that, although the pure state $W = P_{c_1\psi_1 + c_2\psi_2}$, and the mixed state $W' = |c_1|^2 P_{\psi_1} + |c_2|^2 P_{\psi_2}$ yield the same probabilities for the values of B, viz. $|c_1|^2$ for b_1 and $|c_2|^2$ for b_2, they yield different probabilities for an observable A that is incommensurable with B. Consider such an observable A with eigenvectors ϕ_n and eigenvalues a_n. If we take the state of the ensemble to be represented by the mixture W', then the probability that a measurement of A in state ψ will yield the eigenvalue a_k is the weighted sum of the probabilities pertaining to the subensembles:

$$\text{prob}(\psi; A, a_k) = |c_1|^2 \text{prob}(\psi_1; A, a_k) + |c_2|^2 \text{prob}(\psi_2; A, a_k)$$
$$= |c_1|^2 |(\phi_k, \psi_1)|^2 + |c_2|^2 |(\phi_k, \psi_2)|^2,$$

with ϕ_k the eigenvector corresponding to a_k. The probability specified by the superposition W, however, is:

$$\text{prob}(\psi; A, a_k) = |(\phi_k, \psi)|^2 = |c_1(\phi_k, \psi_1) + c_2(\phi_k, \psi_2)|^2$$
$$= |c_1|^2 |(\phi_k, \psi_1)^2 + |c_2|^2 (\phi_k, \psi_2)|^2 + 2\text{Re}\, c_1 c_2^* (\phi_k, \psi_1)(\phi_k, \psi_2)^*.$$

The probability specified by the superposition is a square of sums rather than a sum of squares, and it contains in addition the so-called 'interference' terms, $2\text{Re}\, c_1 c_2^* (\phi_k, \psi_1)(\phi_k, \psi_2)^*$. Moreover, the experimental data confirm the probability formula specified by the superposition W rather than by the mixture W'. Thus, it is held, a superposition cannot describe a mixed state, and so E cannot be regarded as an ensemble which is a mixture of the pure subensembles E_1, E_2.

This objection, however, is not cogent, for it presupposes that the ensemble E must be taken to be a mixed ensemble with respect to the observable A as well as to B. But this presupposition is not required by the objective-values theory; all that is required is that the ensemble E be a mixed ensemble with respect to B; it is not also required that E be a mixed ensemble with respect to any other observables. The question whether E is a mixed ensemble with respect to observables other than B is open; and a negative answer to it is compatible with the objective-values theory. If E were a mixed ensemble with respect to A, we should expect it to be describable by the mixture W'; but if E is not a mixed ensemble with respect to A, then there is no such expectation, and the mixed ensemble E may well be described by the superposition or the so-called pure statistical operator:

$$W = P_{c_1\psi_1 + c_2\psi_2}.$$

In classical physics, of course, a mixed ensemble is generally taken to be mixed with respect to all the observables of an object that define its classical state. But this conception of an ensemble cannot be expected to hold in quantum physics. Exactly; in quantum physics ensembles are defined only over sets of commensurable or compatible observables, and a mixed ensemble is mixed only with respect to such a set. The projection operator W' cannot be regarded as representing a mixture of the quantum-mechanical sort.

Is E a mixed ensemble with respect to A? The fact that the probability function, $\text{prob}(\psi; A, a_k)$, is not correctly specified by the mixture W' suggests that it is not. It can be argued, moreover, that W' specifies the wrong probability function owing to the fact that the expressions, 'prob $(\psi_1; A, a_k)$' and 'prob $(\psi_2; A, a_k)$', are not well defined. There are two main ways of reading these expressions. If, on the one hand, these expressions are taken to mean 'the probability that A has the value a_k in state ψ_1 (or ψ_2)', then the expressions can be held to be not well defined on the grounds that, since ψ_1, ψ_2 are not eigenfunctions of the operator A, A has no definite value in states ψ_1, ψ_2. If, on the other hand, these expressions are taken to mean 'the probability that a measurement of A in state ψ_1 (or ψ_2) will yield the value a_k', then again they are not well defined, since A is indeterminable in these states. Apropos the latter case, if, having measured B, we know that the object is in state ψ_1 (or ψ_2) and we measure A, then the value which we obtain will not pertain to ψ_1 (or ψ_2), since the value that does so pertain, whatever it is, will have been altered by the measurement of B. Thus if a_k were the result of the measurement of A, the state of O would be ϕ_k, or $W'' = P_{\phi_k}$; but W'' would not describe the same ensemble as that described by W. The value of A in states ψ_1, ψ_2 is, therefore, indeterminable. On the objective-values theory the presumption is, of course, that there is no numerical difference between these two interpretations of the probability function, since the value obtained by measurement is the value that the observable possessed immediately before the measurement. Nevertheless the two readings differ in their consequences for the interpretation of the theory; and besides, the reading in terms of measurement accords with Bohr's insistence that an assignment of a state vector is well defined only if it refers to a well-defined experimental procedure. Thus he writes:

> . . . all unambiguous interpretation of the quantum mechanical formalism involves the fixation of the external conditions, defining the initial state of the atomic system concerned and the character of the possible predictions as regards subsequent observable properties of that system.[37]

An analogous difficulty arises in the interpretation of the double-slit experiment. It is widely held that a collection of electrons which are passed, one at a time, through a double-slit diaphragm cannot be said to belong to a mixed ensemble E which is made up of two pure subensembles E_1, E_2 containing the electrons which pass exclusively through slit 1 and slit 2 respectively. For, if E were a mixed ensemble, it should be describable by the mixture $W' = |c_1|^2 P_{\psi_1} + |c_2| P_{\psi_2}$, and not by the putative pure state $W = P_\psi$, where $\psi = c_1\psi_1 + c_2\psi_2$, and ψ_1, ψ_2 describe the states of electrons passing exclusively through slits 1 and 2 respectively. But, while W and W' give rise to the same probability function for some observables, such as 'passage through slit 1', they yield different functions for others, such as 'arrival at position y on the screen'; and in the latter case it is the probability specified by the state W that is confirmed by experiment. Hence the state W, which is a superposition, cannot be said to represent a mixed ensemble, and consequently each electron cannot be conceived of as passing exclusively through one slit.

This argument, however, is far from cogent. It rests upon the premiss that E is a mixed ensemble only if it is describable by the mixed state W'. But this premiss is equivalent to the assumption that E is a mixed ensemble only if the subensembles E_1 and E_2 are the same as the ensembles E', E'' determined respectively by the electrons which pass through slit 1 while slit 2 is closed, and by those which pass through slit 2 while slit 1 is closed. There is, however, no compelling reason to accept this assumption. The state of an electron may differ depending upon whether just one slit or two slits are open when it traverses the diaphragm. If the state does so differ, then the subensembles E_1, E_2 are not the same as E', E'': the fact that E is not describable by W', then, does not entail that E is not a mixed ensemble. E *may* be a mixed ensemble; and if it is, then its state is described by the superposition W. The curious consequence of this interpretation – that the open slit through which an electron does not pass has a causal effect upon its state – is no more curious than the paradoxical consequence of the rival interpretation, namely that each electron passes through both slits. Moreover, it squares with Bohr's doctrine that the state of an object is determined by the *entire* experimental arrangement; it coheres, that is, with the doctrine of *wholeness*. The quantum-mechanical mixture W takes Bohrian wholeness into account, whereas the classical mixture W' does not; indeed, this is the significance of the so-called interference terms in the probability expression generated by W.

An elaboration of the first objection may occur to some: it is that the objective-values theory entails that the subensembles E_1, E_2 are dispersion-free as well as pure, i.e. each observable has the same value in the ensemble (or more precisely, for all observables A, values a, and states W, prob$(W, A, a) = 1$ or 0); but this consequence contradicts the famous corollary to A.M. Gleason's theorem to the effect that there are no statistical operators which are dispersion-free. This further objection fails, however, since the notion of a dispersion-free operator with which the corollary to Gleason's theorem is concerned is one which is defined with respect to *all* the observables of an object, and not just with respect to *one* observable or one set of commensurable observables. All that the objective-values theory requires is that there should be statistical operators which are both pure and dispersion-free with respect to each observable taken individually; and this requirement is satisfied.

The first objection against the objective-values theory, I conclude, fails because it is question-begging. It can be argued, however, that the objective-values theory, together with some plausible assumptions, entails that all the observables of an object have definite values at any time – the intrinsic-values theory of properties. If, the argument goes, we measure an observable A at time t and the result is a_k, then a_k is the value that A possessed immediately before *t*. If, however, instead of measuring *A* at *t*, we had measured an observable *B* which is incommensurable with *A*, then whatever value we had found would have been the value of *B* immediately before *t*. Moreover, we are free to measure at *t* either *A* or *B*, but not both. It is plausible to suggest, therefore, that immediately before *t* both *A* and *B* possessed the values that would have been obtained by measurement, had a measurement been made; both values must, as it were, have been there to be revealed by whatever measurement we chose to make. If this were the case, then all the observables of an object would have definite values at all times, and not simply at a time when they happen to be measured. There can be no doubt that Bohr rejected the intrinsic-values theory. But how could he consistently do this if he also held the objective-values theory? I shall answer this question in the following chapter.

A further possible difficulty for the objective-values theory may arise from the work of J.S. Bell concerning the quantum-mechanical statistics governing the spin values of spin-$\frac{1}{2}$ particles. I shall, however, postpone examining this question until Bell's work is discussed in Chapter 9.

7

Bohr's
theory of properties

In Bohr's view, what ontological significance does measurement have in quantum physics? How, exactly, are we to conceive of the physical properties which we attribute to microphysical objects on the basis of measurement? As we have seen, Bohr regards such properties as objective in the sense that in general they characterise the object immediately before the measurement. But what exactly is their ontological status? In order to answer this question, we shall have to consider in detail Bohr's view of the ontological significance of kinematic-dynamic complementarity. It is generally believed that Bohr held that in quantum physics an object cannot meaningfully be said to possess simultaneously an exact position and an exact momentum. Did he, and if so, why did he?

7.1
The interactive-properties theory

Some hold that the process of measurement does not so much create the value of the measured observable as disturb a pre-existing value. On this view the measured value characterises the observable only *after* the measurement, and not before: the measured value may be said to be a 'created' or 'disturbed' value.[1]

Bohr, as I have already said, seems to have toyed with this view in his earlier years: even as late as 1933 he says:

> Indeed, even in a position or momentum measurement on the electron in a hydrogen atom in a given stationary state one can assert with a certain right that the measuring result is produced only by the measurement itself.[2]

After 1935, however, he expressly rejected the creation-by-measurement view, and hence the interactive-properties theory: talk of 'creating physical attributes to objects by measuring processes' is simply not well defined, and a source of confusion.[3]

Bohr occasionally says that the results of measurement cannot be interpreted as providing 'information about independent properties of the objects'.[4] This statement, however, need not be taken to mean that quantum-mechanical properties are interactive. The more plausible interpretation is that they are not absolute properties, but relative properties, dependent on a frame of reference. Elsewhere he speaks of 'the unavoidable renunciation as regards the absolute significance of ordinary attributes of objects'.[5] Besides, in this context Bohr almost always speaks of *behaviour* rather than of *properties*:

> . . . the unavoidable interaction between the objects and the measuring instruments sets an absolute limit to the possibility of speaking of a behaviour of atomic objects which is independent of the means of observation.[6]

Moreover, in the manuscript of one of his published papers the words 'independent properties' are scored out in two places, and 'independent behaviour' written above.[7] By 'independent behaviour' and 'autonomous behaviour' Bohr means that the behaviour of the object that we observe (e.g. the trajectory) is ineluctably a product of the process of observation, and hence is not autonomous, being different from what it would be if it were not observed. The observable *properties* (synchronic properties) of an object may be non-interactive and objective in the sense that they are not artefacts of the process of observation, even although the observable *behaviour* (diachronic properties) of the object is not objective in that sense.

If quantum-mechanical observables were interactive properties, we should expect that their measured values would in general characterise the object immediately *after* the measurement, and not immediately *before*. But, as we have seen, for most types of measurements, the measured value characterises the observable before, and not after, the measurement. Thus it is only in the case of state preparations that the "measured" value does not characterise the object before the measurement; and state preparations are measurements only in a Pickwickian sense. Besides, there is no convincing textual evidence that Bohr held the interactive-properties theory as one of his settled views.

7.2
The dispositional-properties theory

A different version of the creation-by-measurement theory has been proposed by Margenau, who maintains that certain quantum-mechanical attributes, such as position and momentum, are latent in the sense that they emerge only in response to observation. He distinguishes between 'possessed' properties, such as mass and charge, and 'latent' properties, such as position, momentum, energy and spin. Position is held to be a latent property in the sense that an object in an eigenstate of the momentum operator does not *have* a position; it has a position only when a measurement 'elicits' or 'actualises' it.[8] Margenau does not make clear what this process of 'eliciting' is thought to be, though it seems that he has in mind something like the realisation of a dispositional property. But this leaves the measurement process utterly obscure: latent properties 'in a certain obscure but primitive sense, "come about" or "spring into being" when a measurement is made'.[9] Margenau is unable to explain why it is that an object that is nowhere in particular suddenly comes to be somewhere in particular when an observation of it is made.

Margenau's main reason for proposing a dispositional theory appears to be that an electron's 'ability to interfere with itself clearly implies that "it is in many places at once".'(ib.) In the two-slit experiment, an electron has no definite position before it hits the screen, but a definite position is 'actualised' when it hits. The suggestion that an electron can 'interfere with itself' implies that an electron is an undulatory entity; thus Margenau's grounds appear to be little different from the naive doctrine of wave-particle duality which he purports to reject.

Heisenberg also espouses a dispositional theory that is just as obscure as Margenau's. He holds that unobserved atoms and electrons 'form a world of potentialities or possibilities rather than one of things or facts'.[10] Bub actually attributes the dispositional-properties theory to Bohr: quantum-mechanical observables, in Bohr's view, represent the dispositions of the object 'to behave (i.e. be "disturbed") in certain ways in situations defined by macroscopic (classical) systems'.[11] On this reading, wholeness is the inseparability of the microphysical object from the measuring instrument defining the conditions under which the dispositions are realised; and measurement is not to be regarded as a procedure for establishing whether an object has a certain property, but as a disturbance of a certain kind, which actualises the property.[12] Bub is confounding the dispositional-properties theory with the interactive-properties theory.

It might be thought that the dispositional-properties theory could be made more precise by utilising the propensity theory of probability. The exact position of a microphysical object might be conceived of not as an occurrent property but as a dispositional property that corresponds to a probability that the object will be found at a certain exact position if a position measurement is made, the probability being grounded in an objective propensity of the object, or of the object and an experimental arrangement, to behave in a certain way if certain conditions obtain. But talk of 'propensity' here is simply a surrogate for talk of 'dispositions', and has little explanatory force; what is wanted is a detailed theory of what these propensities or dispositions consist in. An appeal to the propensity theory of probability provides only the mere beginnings of such a theory. As it stands, Margenau's and Heisenberg's view is impenetrably obscure.

I do not believe that Bohr held a dispositional-properties theory of the obscure Heisenberg–Margenau kind. Nevertheless, as we shall see below, there is a sense in which Bohr's theory, when properly construed, is not wholly dissimilar to a dispositional-properties theory.

7.3
The relational-properties theory

Feyerabend rejects the interactive-properties interpretation in favour of a relational-properties account. He holds that quantum-mechanical states, and in particular observable properties, are relational, 'relations between the systems and measuring devices'.[13] Rosenfeld also supports a relational interpretation: the observables of position and momentum 'express *relations* between the system and certain apparatus of entirely classical (i.e. directly controllable) character which serve to fix the conditions of observation and register the results'.[14]

Feyerabend argues that since microphysical attributes are relations between the object and the measuring instrument, a change in the state of the instrument automatically brings about a change in the state of the object, just as the relational attribute 'being larger than *B*' may be changed by lengthening *B*. An interaction between the object and the instrument, then, is not a necessary condition of a change in the properties of the object. On this view the state vector describes not the intrinsic physical state of an object, but its relational state.[15]

Feyerabend makes no attempt to show in detail that Bohr adopted a relational theory of properties. Nevertheless, the relational-properties theory is a very plausible construal of Bohr's position. It coheres well with his insistence that a quantum-mechanical description of an object is well defined only with respect to a specific experimental arrangement, since it is the latter that defines the frame of reference:

> The essential lesson of the analysis of measurements in quantum theory is thus the emphasis on the necessity, in the account of the phenomena, of taking the whole experimental arrangement into consideration, in complete conformity with the fact that all unambiguous interpretation of the quantum mechanical formalism involves the fixation of the external conditions, defining the initial state of the atomic system concerned and the character of the possible predictions as regards subsequent observable properties of that system.[16]

It is for this reason too that Bohr emphasises that the attribution of physical properties is well defined only in relation to some well-defined experimental arrangement specifying a suitable reference frame.[17]

In fact Bohr's notion of *wholeness* has two distinct senses, one dynamic, the other conceptual. On the one hand object and apparatus form a dynamic whole in that during the measurement interaction they are not dynamically separable, since the interaction between them cannot be analysed. On the other hand object and apparatus form a contextual whole in the sense that the object's possessing certain properties is dependent upon the frame of reference defined by the experimental arrangement.

It is possible, however, that Bohr regards kinematic and dynamic properties as relative rather than relational. The property 'tall', for example, is relative, as distinct from absolute, but it is not relational; by contrast, the property 'taller than x' is relational. Doubtless the property 'tall' is instantiated only if the property 'taller than x' is instantiated, but this fact does not make 'tall' a relational property. Bohr's thesis may be that quantum-mechanical properties such as position and momentum are objective, but not absolute properties in the sense that they characterise the object independently of any frame of reference; they are relative to an appropriate frame of reference which is selected by the observer. On this view, a microphysical object could not meaningfully be said to possess a well-defined position, say, independently of some chosen spatial reference frame. This point is perhaps what Bohr has in mind when he speaks of 'an essential element of ambiguity in ascribing conventional physical attributes to atomic objects' and of 'the difficulties in talking about properties of objects independent of the conditions of observation'.[18] Feyerabend does not consider the relative-properties interpretation.

Which of these two interpretations, if either, is correct – the relational-properties, or the relative-properties interpretation? This question cannot properly be answered until we have considered the question of the ontological significance of kinematic-dynamic complementarity.

7.4
The positivist argument for the indefinability thesis

Bohr is generally taken to have held, not only that the exact simultaneous position and momentum of an object cannot be measured, but also that an object cannot meaningfully be said to possess exact simultaneous values of these observables. I shall call the latter view the *indefinability thesis*. There can be little doubt that Bohr held such a view. In the late twenties, for example, he says that the uncertainty relations set a limit not only to 'the *extent* of the information obtainable by measurements, but . . . also . . . a limit to the *meaning* which we may attribute to such information'.[19] Speaking of the epistemic construal of the uncertainty relations, he writes:

> According to such a formulation it would appear as though we had to do with some arbitrary renunciation of the measurement of either the one or the other of the two well-defined attributes of the object . . . [whereas] the whole situation in atomic physics deprives of all meaning such inherent attributes as the idealizations of classical physics would ascribe to the object.[20]

There is, however, no general agreement about what Bohr's grounds for holding this view were.

The most widely held view has been that Bohr's philosophy of physics is a species of radical empiricism or positivism. On this interpretation the indefinability thesis rests upon a positivistic theory of the meaning of the sentences which ascribe a physical property to an object. According to logical positivism, a statement is cognitively meaningful – purports to state a fact – if and only if its truth-value can be determined or confirmed by observation. On this view, since a sentence which ascribes an exact simultaneous position and momentum to an object cannot be conclusively verified, or confirmed, by sensory experience, it is cognitively meaningless, having no genuine descriptive content.

The positivistic interpretation of Bohr's philosophy has been strongly advocated by Bunge.[21] There can be no denying that some of Bohr's statements can plausibly be given a positivistic reading, at least when taken in isolation. But I know of no unequivocal statement of a positivistic character in the published writings. Nevertheless, there is at least one piece of documentary evidence which seems to confirm the positivist reading. In a letter of 9 January 1936 to Bohr, Philipp Frank, a member of the Vienna Circle, expressed the view that Bohr's conception of physical reality was 'positivistic', while Einstein's was 'metaphysical'. In his reply of 14 January 1936 Bohr writes:

> I am very glad to hear from your kind letter that you have given such care to the papers of Einstein and myself concerning the question of reality. I also think that you have caught the sense of my efforts very well.[22]

It is significant that Bohr does not disavow Frank's positivist construal of his position. While this evidence is undoubtedly important, it does not conclusively establish the positivist interpretation. Frank does not spell out precisely what he means by 'positivist': the contrast between 'positivist' and 'metaphysical' that Frank draws in his letter is vague enough to be acceptable to Bohr even though, as I believe, he was not committed to the logical positivist theory of meaning; besides, Bohr had no interest in philosophical labels. I shall return to this letter below.

7.5

The ontic argument for the indefinability thesis

Bunge's positivist interpretation of Bohr's philosophy has been severely criticised by Adolf Grünbaum, who calls it 'a serious error'. Grünbaum maintains that Bohr's reasons for the indefinability thesis stem not from a general philosophical doctrine but from his understanding of the physics. In Bohr's view, exact position and exact momentum are not jointly measurable properties since 'they do not jointly exist to begin with': the property of exact simultaneous position and momentum is simply not well defined.[23]

Grünbaum does not support his interpretation with detailed textual evidence, but the evidence for it is not difficult to find. In the Como paper Bohr talks frequently of the 'definition' of physical properties, contrasting this with 'observation'. He states, for example, that the uncertainty relations express a general reciprocal relation between 'the sharpness of definition of the space-time and energy-momentum vectors associated with the individuals', and also that, with regard to complementarity, we must keep in mind 'the possibilities of definition as well as of observation'.[24] Moreover, whereas observation has to do with the 'space–time co-ordination', definition has to do with the 'claim of causality', i.e. with energy and momentum.[25] The point of the distinction between observation and definition seems to be that when the position of an object, or the time-of-occurrence of an event, has been precisely measured, the object has no well-defined momentum, and there is no well-defined energy associated with the event in question.

Bohr's talk of 'definition' is not unambiguous: it is possible that by 'define' he means 'ascertain', and that by 'definition' he means simply 'measurement', employing the term as a stylistic variant when referring to dynamic properties. While this reading is possible, it is highly unlikely, and I shall not discuss it further.

Why, then, does Bohr regard the property of exact simultaneous position and momentum as being not well defined? This view, according to Grünbaum, is necessitated by experimental data. The double-slit experiment, for example, provides evidence contradicting the assumption that an electron is the kind of entity that possesses an exact though unknown simultaneous position and momentum; for, if the electron were such an entity, it would have a definite, though unknown, trajectory partly determined by the slit through which it does not pass. But we expect the trajectory of a particle to be independent of the slit through which it does not pass.[26] This argument is similar to the one which Bohr employs in the Como paper in order to derive the uncertainty relations; there he suggests that the uncertainty relations are a consequence of the viability of the wave model of the electron. Expressed in this way, the argument seems to be that an electron cannot be said to have an exact simultaneous position and momentum because it is a wave-like entity. But this is not what Bohr has in mind: as we have seen, he holds that the wave model of the electron has no realistic significance. Yet this in turn is not quite correct: the model has *some* realistic significance, to the extent that the electron cannot unambiguously be said to be a classical particle. The point of the argument, then, is, not that an electron does not have a well-defined position and momentum since it is a wave-like entity, but rather that, since the electron cannot unambiguously be said to be a particle, it cannot unambiguously be said to have simultaneously an exact position and an exact momentum. This is what Bohr means when he suggests that we replace the idea of 'a coincidence of well-defined events in a space–time point' with the notion of 'unsharply defined individuals within finite space–time regions'.[27] Grünbaum is aware of this important qualification to the argument, though he does not bring it out, relegating it instead to a footnote.[28] The real significance of wave-particle complementarity for Bohr is that the classical notion of a particle as an entity with an exact simultaneous position and momentum, and hence also an exact trajectory, has no application in quantum physics. Quantum-mechanical particles are not wave-like entities, 'wavicles' as A.E. Eddington aptly called them; rather, the classical sense of the term 'particle' is generalised in quantum mechanics, in that certain entities are called 'particles' in quantum mechanics that would not be entitled to that designation in classical mechanics.

Bohr's disagreement with Heisenberg over the derivation of the uncertainty relations becomes much clearer in the light of this interpretation. As we have seen, Heisenberg believed that wave-particle duality had no role to play in the derivation of the uncertainty relations. Like Bohr, he was inclined to give these relations an ontic as well as an epistemic interpretation; for Heisenberg, however, the ontic significance of these relations stems not from physical considerations but from a general philosophical outlook, viz. positivism.[29] For Bohr, by contrast, the ontic significance of the uncertainty relations is grounded primarily in the physics: these relations determine 'the highest possible accuracy in the definition of the energy and momentum of the individuals associated with the wave-field'.[30] In order to bring out the full significance of the uncertainty relations, he says, 'a closer investigation of the possibilities of definition would still seem necessary'.[31] The ontic significance of wave-particle complementarity is the breakdown of the universal applicability of the classical models; the uncertainty relations are a precise mathematical expression of that breakdown.

We see here the close connection between Bohr's thinking in 1927 and his response to the failure of the Bohr–Kramers–Slater theory in 1925. Just before the invention of quantum mechanics Bohr had stressed that the quantum theory did not permit a space-time description of objects in terms of visualisable pictures. The failure of classical electrodynamics necessitated 'a thoroughgoing revolution in the concepts upon which the description of nature has rested up to now'.[32] Bohr, however, did not regard the failure of the classical models as complete. Wave-particle complementarity indicated that the classical models could be employed in restricted contexts; yet kinematic-dynamic complementarity indicated that even in these restricted contexts the classical models could not be given their full classical sense. It is thus not only the classical concept of observation (entailing no disturbance of the object) that is an idealisation, but also the classical concept of definition, that is, the idea that objects have at all times a well-defined momentum and a well-defined energy. As Bohr puts it, the space-time co-ordination and the claim of causality symbolise 'the idealization of observation and definition respectively'.[33]

Bohr firmly rejected Schrödinger's realist interpretation of wave mechanics as a theory of matter waves; yet he regarded wave mechanics as a major theoretical advance. What, for Bohr, was the significance of the wave function and the superposition principle in wave mechanics? In the manuscript papers of 1927 Bohr frequently speaks of the complementarity of 'individuality and superposition' rather than of particle and wave. Since 'the wave-mechanical solutions can be visualized only in so far as they can be described with the aid of the concept of free particles',[34] the superposition of wave functions is not to be understood as representing the superimposition of real physical waves, but rather as a mathematical representation of the breakdown of the classical model of a particle.

Yet, if the substantive content of wave-particle complementarity is simply the indefinability thesis, on what is that thesis itself based? Not, obviously, on wave-particle complementarity. Grünbaum suggests a second ground for it, viz. that quantum-mechanical properties are interactional.[35] But I have already rejected that view. Feyerabend suggests a different ground: like Grünbaum, he maintains that Bohr's main reasons for the indefinability thesis were not primarily philosophical, and in particular not positivistic, but physical. To paraphrase Feyerabend, Bohr's main argument is that, owing to the quantum postulate, interactions which involve an exchange of energy are discontinuous in the sense that the energy states of the interacting objects change discontinuously from one discrete state to another; and hence, when the interaction takes place during a finite time interval t, none of the interacting objects can be in a definite energy state (though of course the system comprised of the interacting objects will be in a definite energy state). Now, since observation or measurement generally involves a dynamic interaction between the object and the instrument of observation, neither the object nor the instrument can be said to be in any definite energy state during the interval t of the measurement interaction. Thus, when the position of an object is measured, the measurement process renders the energy of the object indefinite at t (and the energy of the instrument as well). The process also renders the momentum of the object indefinite, since energy is definable in terms of momentum. Similarly, when the energy or momentum of an object is measured, the object has no definite position during the interval t when the object and the instrument are interacting.[36]

Some of Bohr's statements may plausibly be taken to suggest that he sometimes has some such argument in mind. He says, for example, that if the time interval t during which objects interact is shorter than the vibration period associated with the transition process, then the interacting objects do not possess definite energies at t.[37] This is the point that he has in mind when he says, apropos the measurement interaction, that 'an independent reality in the ordinary physical sense can neither be ascribed to the phenomena nor to the agencies of observation'.[38] The meaning of this statement is that, during the interaction between them, neither the object nor the instrument can be said to be in a definite dynamic state. It is what he has in mind also when he states that '*the finite magnitude of the quantum of action prevents altogether a sharp distinction being made between a phenomenon and the agency by which it is observed*'.[39] Since, during their interaction, neither the object nor the instrument is in a definite energy state, the two form a whole in which no sharp distinction between the component parts can be made.

The argument which Feyerabend attributes to Bohr, however, shows only that an object does not have an exact energy or momentum during the interval t in which it is interacting with another object; it does not show that an object does not have an exact position and momentum at t. If, however, one adopts the ontic interpretation of the state vector, there is reason to think that the indefiniteness in the state vector of the object during its interaction with another object persists after the interaction has ceased, for according to quantum mechanics, after the two objects have interacted, the state vector of the system comprising them is no longer the product of the separate vectors belonging to the component objects, and hence neither object can be assigned a state vector of its own. Moreover, the indefiniteness in the state vectors of the component objects persists until one or other of them interacts with some further object. I do not believe, however, that this argument does constitute Bohr's grounds for holding the indefinability thesis, for there is no reason to think that he adopted the ontic interpretation of the state vector. Certainly the fact that an object is in no definite dynamic state while it is interacting with another object was regarded by Bohr as being extremely important; but Bohr did not think that this fact alone entailed the indefinability thesis. This fact entails the indefinability thesis only when taken in conjunction with certain philosophical considerations; it was these which constituted Bohr's main grounds for the thesis.

7.6
The semantic argument for the indefinability thesis

After 1935 Bohr expressed the indefinability thesis in what may be called 'semantic', as distinct from 'ontic', terms, or in 'formal' rather than 'material', terms. For example:

> Thus, a sentence like "we cannot know both the momentum and the position of an object" raises at once questions as to the physical reality of two such attributes of the object, which can be answered only by referring to the conditions for the unambiguous use of space-time concepts, on the one hand, and dynamical conservation laws, on the other hand.[40]

Thus construed, complementarity does not so much express the ontic mutual exclusiveness of the properties of exact position and exact momentum as convey a limitation on the meaningful applicability of the respective concepts. Bohr's interest is not so much ontological as semantical: he is concerned not so much with the nature of physical objects as with the character of the concepts under which we subsume objects. Bohr's central point is that

> ... a consistent application even of the most elementary concepts indispensable for the description of our daily experience, is based on assumptions initially unnoticed, the explicit consideration of which is, however, essential if we wish to obtain a classification of more extended domains of experience ... the analysis of new experiences is liable to disclose again and again the unrecognized presuppositions for an unambiguous use of our most simple concepts, such as space-time description and causal connection.[41]

The point is that the meaningful applicability of a concept may presuppose certain conditions (a context) which may not be explicitly recognised since it is tacitly assumed, perhaps wrongly, that they are universally satisfied. Bohr likens complementarity to relativity theory in this respect: both theories are concerned with the 'foundation for the unambiguous use of elementary physical ideas'.[42] Just as it was the hypothesis of the finite and constant velocity of radiation that led Einstein to argue that the necessary preconditions for the meaningful applicability of the classical concepts of absolute space, time and simultaneity do not obtain, so the discovery of the finite and constant quantum of action led Bohr to argue that the preconditions for the meaningful applicability of the classical notion of exact simultaneous position and momentum are not satisfied. Thus:

In both cases we are concerned with the recognition of physical laws which lie outside the domain of our ordinary experience and which present difficulties to our accustomed forms of perception. We learn that these forms of perception are *idealizations*, the suitability of which for reducing our ordinary sense impressions to order depends upon the practically infinite velocity of light and upon the smallness of the quantum of action.[43]

Whereas relativity undermines the notion of an unambiguous separation between space and time without reference to the observer, so complementarity undermines the notion of a sharp separation between the behaviour of objects and their interaction with the instruments of observation.[44] Both theories disclose 'the essential dependence of every physical phenomenon on the standpoint of the observer',[45] and undermine the notion of 'the absolute significance of conventional physical attributes of objects'.[46]

The analogy which Bohr drew between the theory of relativity and his theory of complementarity may not have been a mere afterthought; it may have played a formative role in his thinking at an early stage. The analogy is referred to in the Como article, and it is touched upon in his correspondence of 1927.[47] What struck Bohr was the fact that the quantum postulate and the postulate of the finite velocity of electromagnetic radiation both undermine important presuppositions of classical physics. This fact, he was convinced, entailed significant changes in the logical structure of the classical conceptual scheme: the main philosophical message of relativity theory – that what one can significantly say about an object depends upon one's spatio-temporal frame of reference – applies to quantum theory just as much as it does to relativity theory. Thus kinematic and dynamic descriptions of an object must be made relative to specific experimental arrangements – specific pieces of hardware – in much the same way as attributions of position and velocity, length etc., in relativity theory are made relative to specific co-ordinate systems. Seeing things in this way, Bohr was naturally disappointed that Einstein did not share his view of the analogy which he saw between the two great physical theories of the twentieth century.

Bohr's concern in the philosophy of physics was always primarily with conceptual issues: even before the advent of quantum mechanics his interest was mainly in the implications of the notion of discontinuity for the classical concepts of particle and wave. His concern with the logical foundations of our application of concepts thus antedates his introduction of the idea of complementarity.

Bohr continually stresses that the quantum of action necessitates a 'radical revision of the foundation for the unambiguous use of our most elementary physical concepts'.[48] The revision required is a generalisation of the classical conceptual scheme that will enable it to comprise the notion of indivisibility and thereby a wider domain of experience.[49] The notion of generalisation is as fundamental to Bohr's theory of complementarity as it is to his theory of correspondence. He writes:

> The main point to realize is that all knowledge presents itself within a conceptual framework adapted to account for previous experience and that any such frame may prove too narrow to comprehend new experiences.[50]

If a conceptual scheme which has proved adequate over a wide range of empirical data fails in a new area of experience, it need not be discarded; rather, it may be modified through generalisation – through a weakening of its fundamental presuppositions. A conceptual scheme may fail not because its conceptual structure is too poor, but because it is too rich. Bohr describes his philosophical attitude as:

> ... the endeavour to achieve a harmonious comprehension of ever wider aspects of our situation, recognising that no experience is definable without a logical frame and that any apparent disharmony can be removed only by an appropriate widening of the conceptual framework.[51]

The point of the semantic argument is that the properties of exact position and exact momentum cannot be meaningfully ascribed simultaneously to one and the same object, since the preconditions for the meaningful ascribability of these properties are mutually exclusive. But what exactly are these preconditions?

7.7
The substance of the semantic argument

The preconditions for the meaningful ascribability of the physical concepts are determined by the experimental arrangement. Thus Bohr writes:

> Indeed we have in each experimental arrangement suited for the study of proper quantum phenomena not merely to do with an ignorance of the value of certain physical quantities, but with the impossibility of defining these quantities in an unambiguous way.[52]

The experimental arrangement, moreover, is a *measuring* instrument: 'unambiguous application of the concepts used in the description of phenomena depends essentially on the conditions of observation'.[53] The presence of an appropriate measuring instrument is, then, a necessary condition of the meaningful applicability of the physical concepts. This condition is a consequence of Bohr's thesis that the physical properties of position and momentum are relative rather than absolute, relative to a frame of reference defined by some macroscopic body.

Is the presence of an appropriate measuring instrument also a sufficient, as well as a necessary, condition of meaningful ascribability, so that an electron could be said to have a definite, though perhaps unknown, position provided a suitable position measuring instrument were to hand? It would simplify matters if this were the case; but alas it seems that it is not, for the conditions of meaningful ascribability are determined not merely by a measuring instrument but also by a phenomenon in Bohr's technical sense. This being the case, the performance of a well-defined *measurement*, in addition to the presence of a well-defined *instrument*, is required. I shall call the former of these two conditions the 'strong meaning condition', and the latter the 'weak meaning condition'. That Bohr adopts the strong meaning condition is made clear in his important Warsaw lecture:

> . . . all unambiguous interpretation of the quantum mechanical formalism involves the fixation of the experimental conditions, defining the initial state of the atomic system concerned and the character of the possible predictions as regards subsequent observable properties of that system. Any measurement in quantum theory can in fact only refer to a fixation of the initial state or to the test of such predictions, and it is first the combination of measurements of both kinds which constitutes a well-defined phenomenon.[54]

By adopting the strong meaning condition, Bohr may seem to be espousing the positivist argument. That argument is indeed, *prima facie*, a plausible interpretation. But I have already rejected that interpretation; so I shall not pursue the point. By the same token, however, Bohr may seem to be adopting an operationalist theory of the meaning of physical predicates. Since operationalism is a distinct doctrine from radical empiricism or logical positivism, the operationalist interpretation must be seriously examined.

According to operationalism the meaning of a physical concept is defined by the physical operation which is required in order to tell whether an object falls under the concept, or by the operation which constitutes a measurement of the property which the concept expresses. Operationalism was very much in the air in the twenties, and it was brought to the fore in 1927 – the year of complementarity – by the publication of P.W. Bridgman's *The Logic of Modern Physics*. There can be little doubt that Bohr was aware of the operationalist current in the air, and it is not unlikely that he was to some extent influenced by it.

There can be no denying that Bohr's argument for the indefinability thesis is operationalist at least to the extent that he agrees with the classical operationalist that a physical predicate is meaningfully applicable only in a context which includes a measurement of the property which the predicate expresses. This, of course, is not to say that Bohr is an operationalist of the classical, Bridgmanian sort, for he does not hold that the measurement operation exhausts the meaning of the predicate, but merely that it constitutes a necessary condition of its meaningful applicability. Moreover, Bohr did not set out with any commitment to the operationalist doctrine as a philosophical programme; that sort of commitment was quite alien to his way of thinking. It would be wrong, then, to say that Bohr's theory of meaning is operationalist *sans phrase*. It is legitimate to speak, with appropriate qualification, only of an operationalist ingredient in Bohr's theory of meaning; but the term 'operationalist' is perhaps best avoided altogether, as being more misleading than informative. If Bohr's theory of meaning was operationalist, it would not provide an adequate basis for his indefinability thesis, since operationalism as a general theory of meaning has been wholly discredited.

Why exactly did Bohr impose the strong meaning condition, in addition to the weak one? His main reason, I believe, was that he thought that the weak condition was not in itself sufficient to circumvent the objections which were raised against the indefinability thesis by Einstein, Podolsky and Rosen. Certain philosophical considerations, moreover, predisposed him in favour of the strong condition; I shall discuss these in Chapter 10.

7.8
Difficulties with the strong meaning condition

Can Bohr consistently hold both the objective-values theory and the strong meaning condition? The objective-values theory contains the idea of the pre-existing value of an observable; but how, given the strong meaning condition, can an observable meaningfully be said to have a well-defined pre-existing value immediately *before* it has been measured? The difficulty is real if Bohr adopted the ontic interpretation of the state vector, for on this interpretation, once they have interacted, the object and the instrument do not have definite states of their own, since the state vector of the system comprising them cannot be factorised into distinct vectors belonging individually to the object and the instrument; and the indefiniteness in their states continues until one of them interacts with some further object. If Bohr adopted the ontic interpretation, then indeed, he could not consistently have held the objective-values theory, since on this ontic interpretation an observable does not have a definite value immediately before it is measured. Might not this interpretation have constituted Bohr's grounds for the strong meaning condition? It is possible that it did; in the Warsaw lecture he states:

> . . . no well-defined use of the concept of "state" can be made as referring to the object separate from the body with which it has been in contact, until the external conditions involved in the definition of this concept are unambiguously fixed by a further suitable control of the auxiliary body.[55]

I do not believe, however, that Bohr did adopt the ontic interpretation of the state vector. The above statement may seem to imply that the notion of the pre-existing state – the state which the object is in immediately before a measurement is made – has no well-defined sense. But when this statement is read in its context – Bohr's discussion of the EPR paper – it is clear that he is intending to say only that if we do not make a measurement on an object, we cannot meaningfully talk about its state, and not that the notion of the pre-existing state of an object is meaningless. Bohr did not intend to say that after the object and the instrument have interacted neither of them is in a definite state until an observation is made on the instrument. He intended to say, rather, that talk of the state of an object is well defined only in the context in which a relevant measurement has been performed. There is, however, no real inconsistency here. The objective-values theory states merely that the *measured* value of an observable is the immediately pre-existing value; and so the required reference to a measurement is satisfied. The theory does not state, or even entail, that an observable has a pre-existing value whether or not it is measured. Bohr, then, can consistently say that successful measurement of an observable reveals its pre-existing value, and also that talk of a pre-existing value is meaningful only in a context in which a measurement has been made. He is, of course, required to deny that we can meaningfully say that the pre-existing value that is revealed by measurement would have been present even if no measurement had been made. He would, however, have been perfectly happy to deny this, since, while maintaining the objective-values theory, he rejected the intrinsic-properties theory.

If Bohr did hold that an object which has interacted with an instrument has no definite state before an observation on the instrument is made, then there is a further difficulty which Feyerabend has pointed out, which is that if an object cannot meaningfully be said to be in a well-defined premeasurement state, it makes no sense to say that it is disturbed by the process of measurement; for in this case there is no well-defined premeasurement state there to be disturbed.[56] Bohr, Feyerabend maintains, was aware of the difficulty, and his response to it was to abandon the notion of the measurement interaction, as the price to be paid for retaining the strong meaning condition. Following Feyerabend, it is now fashionable to say that after 1935 Bohr abjured all talk of disturbance of the object by the measuring instrument. He does indeed deprecate talk of 'disturbing a phenomenon by observation'.[57] I interpret this remark differently, however, as meaning that talk of disturbance by measurement may mislead one into thinking that the result of measurement is not the pre-existing value of the measured observable, but a 'disturbed' value.

Pace Feyerabend, Bohr did not abandon the notion of the measurement disturbance, since that notion is implicit in the concept of the measurement interaction. In order for a measurement to be made an object must act upon an instrument, and consequently react against it, thereby being disturbed. The notion of the interaction plays a crucial role in Bohr's theory of measurement. Even in the EPR experiment the notion plays an indispensable part. Einstein's aim was to devise an experiment in which the object would not be disturbed by a measurement. The experiment which he devised does indeed fulfil that aim; but even here a measurement interaction plays an indispensable role. Bohr recognised this, and he stresses the point in his reply to Einstein:

> Indeed the *finite interaction between object and measuring agencies* conditioned by the very existence of the quantum of action entails – because of the impossibility of controlling the reaction of the object on the measuring instruments if these are to serve their purpose – the necessity of a final renunciation of the classical ideal of causality and a radical revision of our attitude towards the problem of physical reality.[58]

Indeed, he re-iterates this point in two further passages in this paper:

> In fact, the renunciation in each experimental arrangement of the one or the other of two aspects of the description of physical phenomena . . . depends essentially on the impossibility, in the field of quantum theory, of accurately

controlling the reaction of the object on the measuring instruments, i.e. the transfer of momentum in case of position measurements, and the displacement in case of momentum measurements.[59]

Besides, Bohr stresses the notion of the indeterminable measurement interaction again and again in his later writings.[60]

Furthermore, Feyerabend fails to notice that if Bohr did in fact abandon the notion of the measurement interaction after 1935, he was curiously inconsistent in basing the indefinability thesis on the ontic argument which Feyerabend attributes to him, since the notion of the interaction – the exchange of energy between the object and the instrument – is absolutely central to that argument. Bohr, of course, was not being inconsistent: he appreciated very well that if the notion of the measurement interaction is dispensed with there is no way of deriving that crucial component in the concept of kinematic-dynamic complementarity, viz. the idea that results from different measurements cannot be extrapolated to one and the same time. If we dispense with the notion of the interaction, we cannot show what the quantum of *action* has to do with the uncertainty principle, with kinematic-dynamic complementarity, construed either in the epistemic, or the ontic, sense. The notion of the wholeness of the object and the instrument is no substitute for the notion of the interaction, since the notion of wholeness presupposes the notion of the interaction, the quantum of *interaction*, as it were.

7.9
The logic of the semantic argument

Bohr maintains that conjugate kinematic and dynamic properties in quantum mechanics are objective, but also relative to mutually exclusive experimental arrangements and measurement procedures; they are not absolute or intrinsic properties in the sense that objects possess them independently of any contextual conditions, and any ascription of them presupposes that the relevant conditions obtain. On Bohr's view, then, the objective-values theory does not presuppose the intrinsic-properties theory.

What is the logical structure underlying the semantic argument? How is the notion of presupposition to be understood here? The concept of presupposition that Bohr has in mind is similar to the concept introduced by P.F. Strawson in his debate with Bertrand Russell concerning the question of the truth-value of sentences containing non-denoting terms. A sentence S presupposes a sentence T in Strawson's sense if and only if S is true or false only if T is true: if T is false, then S lacks a truth-value.[61] Although Strawson's notion of presupposition is concerned primarily with non-denoting subject expressions, it can be given an application with respect to predicate expressions: the applicability of a predicate F presupposes condition C if and only if any sentence predicating F is true or false only if C obtains. Bohr holds that the notion of exact simultaneous position and momentum has no meaningful application, since the preconditions of the meaningful applicability of the component concepts are mutually exclusive. What Bohr has in mind, I suggest, is that any ascription of this property to an object is meaningless in the sense that it lacks a truth value – is devoid of propositional content.

Bas van Fraassen's notion of supervaluations provides an elegant formal treatment of the notion of a sentence's lacking a truth-value.[62] Karel Lambert has applied the notion of supervaluation to the semantics of quantum-mechanical sentences which ascribe a definite position or momentum to an object – 'experimental sentences' as I shall call them. He does not, however, employ the notion of presupposition in this context, and thinks even that his account of the semantics of these sentences is incompatible with Bohr's approach, on the grounds that, for Bohr, these sentences, when conjoined, are meaningless, not in the semantic sense of lacking a truth-value, but in the syntactic sense of being not well formed.[63] On the contrary, Lambert's semantic approach applies very nicely to Bohr's treatment of these sentences. C.A. Hooker and van Fraassen have provided a semantic theory of Bohr's treatment of experimental sentences of the sort 'Observable A of object O will be found on measurement to have the value a'.[64]

It is important to note that the semantic argument does not require a non-classical logic: the classical propositional calculus continues to hold for all statements which have a truth-value. The theory of supervaluations is simply an extension of the classical calculus that enables it to take account of sentences which lack a truth-value. Bohr himself recognised this point: the resort to a many-valued logic in the interpretation of quantum mechanics is, he held, quite unnecessary; classical logic suffices.[65]

In the light of this semantic approach, it is clear in what sense Bohr holds quantum mechanics to be an indeterministic theory: the theory is not just epistemically indeterministic, but also ontically, or more precisely, semantically indeterministic, in the sense that the concept of exact simultaneous position and momentum is not well defined. It is for this reason that he holds 'discussion about an ultimate determinism or indeterminism of physical events' to be futile.[66]

Bohr's adoption of the strong meaning condition makes the adoption of the objective-values theory difficult to sustain. Since, as I maintain, Bohr was not constrained to adopt that condition, he ought instead to have adopted the weak meaning condition, which, as a condition of the meaningful ascribability of the physical predicates, requires not the actual performance of a measurement but merely the presence of an appropriate measurement arrangement. Given the weak condition, one may, perfectly consistently, hold the objective-values theory and reject the intrinsic-properties theory; for the objective values that are revealed by measurement are not intrinsic or absolute properties in the sense that they characterise the object independently of any experimental arrangement which we may set up. The objective values are relative to the experimental conditions which we set up; once these conditions have been fixed, certain observables possess objective values relative to these conditions.

8

Einstein
versus Bohr

Between 1927 and 1936 no one had a greater influence on Bohr's thinking than Einstein. It was Einstein's unremitting criticism which provided the severest test of Bohr's interpretation of quantum physics, and which forced him to clarify and refine his arguments. The debate between Bohr and Einstein concerning the interpretation of quantum physics is generally recognised as one of the great intellectual disputes in the history of science. What was the nature of the disagreement between them, and what were the arguments with which each defended his position?

8.1
The fifth Solvay Conference, 1927

Einstein was not present at the Volta Centennial Conference at Como in September 1927, where Bohr first presented his theory of complementarity; Einstein first learned of Bohr's interpretation of quantum mechanics at the fifth Solvay Conference of Physics at Brussels in October 1927. In the discussion following Bohr's paper Einstein examined an imaginary experiment in which an electron passes through a slit in a diaphragm and impinges at a point on a hemispherical screen. This experiment, Einstein suggested, could be interpreted in either of two ways: the de Broglie–Schrödinger waves represent either (*a*) an ensemble of electrons spread out in space (but not an individual electron), or (*b*) a wave-packet corresponding to an individual electron.[1] There is, he noted, a difficulty with the latter interpretation. Since the wave-packet produces an effect only at a single point on the screen, and not, as might be expected, over a broad front, there must be a curious action at a distance which brings about the collapse of the packet, preventing it from producing effects at more than one point.[2] For this reason Einstein preferred interpretation (*a*), according to which the de Broglie–Schrödinger waves, and consequently the wave function, represent the state of a statistical ensemble of objects, and not real matter waves, as interpretation (*b*) implies.

Einstein's remarks provide one of the earliest reasonably clear statements of the statistical interpretation of the wave function or state vector, according to which the state vector describes, not the state of an individual object, but the state of a statistical ensemble of objects.[3] What troubled Einstein was the suggestion, already being taken seriously by many, that the probabilistic or statistical character of quantum mechanics is ultimate in the sense that there is no underlying deterministic theory that stands to quantum mechanics as classical mechanics stands to classical statistical mechanics.

During informal discussions with Bohr and Ehrenfest, Einstein criticised Bohr's thesis of kinematic-dynamic complementarity in its epistemic form, and in particular the thesis of the indeterminability of the measurement interaction. He objected that it should be possible to circumvent the uncertainty principle by determining the exact exchange of energy or momentum in individual measurement interactions, and he tried to demonstrate this possibility by means of imaginary experiments. Bohr counter-argued that further analysis of these experiments shows such a possibility to be ruled out.

In his classic account 'Discussion with Einstein on Epistemological Problems in Atomic Physics', Bohr describes the sort of experiments which he and Einstein discussed. He demonstrates how the uncertainty principle is satisfied in the single-slit experiment. If, for example, the diaphragm is fitted with a shutter, the time at which an object passes through it can be determined with arbitrary accuracy. According to the wave model, the shorter the interval Δt during which the slit is open, the shorter the wave-train that passes through the diaphragm, and consequently the greater the spread Δv in the frequencies: $\Delta v \approx 1/\Delta t$, which, in virtue of $E = hv$, gives $\Delta E \Delta t \approx h$. Moreover, the exchange of momentum Δp between the object and the shutter involves an energy exchange,

$$v\Delta p \approx \Delta q \Delta p / \Delta t \approx h/\Delta t,$$

which is of the same order of magnitude as ΔE ('v' signifying the velocity of the shutter). According to the particle model this uncertainty represents a limitation on the possibility of measuring the momentum and energy exchanged between the object and the diaphragm.[4]

Bohr argued that if the spatio-temporal variables in this experiment were exactly controlled, it would be impossible to determine exactly the momentum-energy exchange between the object and the diaphragm.[5] In order to reduce the inaccuracy ΔE it would be necessary to make d/v as small as possible ('d' signifying the height of the slit). But the smaller d/v becomes, the greater the inaccuracy Δt in the knowledge of the time at which the slit is open.

Bohr demonstrated the same point with respect to the double-slit experiment. If we were to measure the exchange of momentum between the object and the diaphragm with great accuracy, the position of the diaphragm would necessarily remain uncertain by an amount specified by the uncertainty relation. The momentum exchanged between the object and the diaphragm would differ, depending on whether the object passed through the upper slit or the lower slit, this difference being of the order of ha/λ ('a' signifying the angle between the paths of the objects passing through the upper and the lower slits, and 'λ' the de Broglie wavelength (see Figure 6)). If this transfer of momentum were to be measured with great accuracy, the position of diaphragm D_1 would necessarily be uncertain by an amount λa. If diaphragm D_2 were mid-way between diaphragm D_1 and the screen S, then the number of 'interference' fringes on the screen per unit length would be $\lambda a/1$. An uncertainty of order λa in the position of D_1 however, would entail an equal uncertainty in the positions of the fringes, and the 'interference' effect would be obliterated.[6]

The most essential point in these considerations is the requirement that in order that the detailed exchange of momentum between the object and the instrument can be measured, some part of the instrument must be freely movable and not rigidly connected to the rest of the apparatus. If for example the time at which an object passes through a diaphragm were to be determined by means of a moving shutter, the shutter would have to be rigidly connected to a robust clock itself connected to the rest of the apparatus which defines the chosen temporal reference frame. But this circumstance would preclude the possibility of measuring the exchange of energy between the object and the shutter.[7]

8.2
The sixth Solvay Conference, 1930

Einstein resumed his attack on the uncertainty principle at the sixth Solvay Conference in 1930. He considered a box with a hole in its side which could be opened and closed by means of a shutter activated by a clock-work mechanism contained within the box. If the box contained a radiation

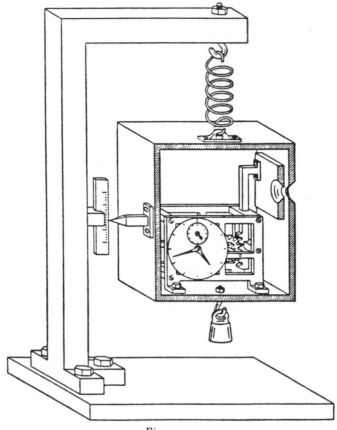

Figure 7

(By courtesy of Prof. P.E. Schilpp.)

source, a single photon could be released from the box by opening the hole for a very short interval at a certain chosen time. If the box were weighed both before and after the release of the photon, the energy of the photon could be determined by recourse to the equation $E = mc^2$ of relativity theory. It would thus be possible to measure with arbitrary accuracy the energy of the photon at the exact time of its release from the box, contrary to the uncertainty relation, $\Delta E \Delta t \geqslant h$ (see Figure 7)[8].

Bohr's ingenious reply hoists Einstein with his own petard. Owing to the highly condensed form in which he presents it, however, Bohr's argument has been widely misunderstood. He argues as follows. If the box were weighed by means of a spring balance, the uncertainty Δm in the measurement of the mass m would be determined by the uncertainty Δy in the reading of the position of the pointer. Owing to the uncertainty relation, $\Delta y \Delta p_y \geqslant h$, the uncertainty Δy would entail an uncertainty Δp_y in the momentum of the box. This uncertainty must obviously be less than the total momentum which the gravitational field can impart to the box during the whole interval T of the balancing operation. Thus,

$$Tg\Delta m > \Delta p_y,$$

where 'g' is the gravitational constant. In virtue of the uncertainty relation we have

$$Tg\Delta m > \hbar / \Delta y$$

and hence

$$Tg\Delta m \Delta y > \hbar.$$

Consequently, the smaller the uncertainty Δy, and hence Δm, the longer the interval T.

Now, a clock which is displaced in the direction of the gravitational force by the amount δT. There is, then, an uncertainty Δt in the rate of the clock: accordance with the 'red-shift' formula of the general theory of relativity:

$$\delta T = Tg\delta y / c^2.$$

Since the position of the box is known only to within Δy, it may be displaced by that amount, and hence the rate of the clock within the box may change by the amount δT. There is, then, an uncertainty Δt in the rate of the clock:

$$\Delta t = \delta T = Tg\Delta y / c^2.$$

Inserting $\Delta y = \Delta t c^2 / Tg$ into the formula, $Tg\Delta m \Delta y > \hbar$, we have

$$Tg\Delta m \Delta T c^2 / Tg > \hbar,$$

which, in view of the equation, $E = mc^2$, gives

$$\Delta E \Delta t > \hbar.$$

Apropos the 'red-shift' formula, the essential point is that as the interval T increases, which it must do if Δy is to be small, so too does the uncertainty Δt. Thus increasing the accuracy of the weighing operation entails increasing the uncertainty concerning the time at which the photon is omitted. It is most important to note that the expression Δt denotes the uncertainty in the rate of the clock, and not the change ΔT in the rate of the clock which occurs before equilibrium is reached in the balancing process – that change is measurable. What concerns Bohr is the possible, but *unmeasurable*, displacement and consequent change in the rate which may occur once the pointer lies within the spatial interval Δy.

Einstein accepted Bohr's argument. Others, however, have rejected it. The most prominent critic is Popper, who objects that if Bohr's interpretation of quantum mechanics requires an appeal to Einstein's theory of gravitation, then this requirement amounts to the strange assumption that the quantum theory contradicts Newton's theory of gravitation, and that Einstein's theory can be derived from the quantum theory.[9] This objection is unconvincing. First, the fact (if it is a fact) that the conjunction of quantum mechanics (according to Bohr's understanding of it) and Newton's theory of gravitation is false, and the conjunction of quantum mechanics with Einstein's theory of gravitation is true, does not imply that quantum mechanics logically entails Einstein's theory.[10] Secondly, theories generally have empirical consequences not *per se* but only in conjunction with other, auxiliary assumptions and theories. Since a weighing process is involved in Einstein's experiment, a theory of gravitation is an immediately relevant auxiliary assumption, and not, as Popper implies, an 'extravagant assumption'.[11] Bohr's adoption of what he sees as the best available theory of gravitation is, then, perfectly legitimate. Moreover, Popper himself admits that resort to a new auxiliary assumption in order to save a theory from refutation is legitimate provided the new assumption is not *ad hoc*. Bohr's resorting to Einstein's gravitational theory, however, is not obviously *ad hoc* in any of the widely used pejorative senses of this term, viz. 'not independently testable', 'lacking independent theoretical support', 'not cohering with the basic concepts and principles of the theory in question', etc. Besides, from the purely polemical point of view, Bohr's assumption satisfies the condition that Popper himself regards as necessary for the legitimate polemical use of imaginary experiments, namely that they be 'concessions to the opponent, or at least acceptable to the opponent'.[12]

Appeal to the general theory of relativity is not in fact necessary for Bohr's argument, since the 'red-shift' formula can be derived from the special theory of relativity together with the principle of the equivalence of inertial and gravitational mass; and Einstein's argument employs the former explicitly and, as Bohr notes, the latter implicitly.[13]

The objection may be made, perhaps, that Bohr ought to have defended the uncertainty principle for energy and time by employing only quantum-mechanical considerations, and in particular the notion of the quantum of action; instead, his defence is based upon the 'red-shift' formula. This objection misses the mark. The uncertainty Δt in the rate of the clock is due primarily to the fact that, owing to the uncertainty principle for position and momentum, the weighing operation requires a comparatively long time. Thus, as Bohr himself remarks, the essential point of the argument has to do with the uncertainty principle and with what it entails in this context.

A word of comment is appropriate at this point concerning the uncertainty relation for energy and time $\Delta E \Delta t \sim h$, since the relation has been the subject of controversy ever since it was introduced by Heisenberg. Heisenberg took the relation to mean that the energy of an object cannot be measured precisely during a short time-interval: the shorter the duration of the measurement operation, the greater the uncertainty in the energy measurement.[14] Bohr did not demur at this interpretation, as is obvious from his reply to Einstein's argument concerning the photon box. But he also interpreted the relation as signifying a relation between the uncertainty in the time at which a process takes place and the uncertainty in the energy transferred during this process.[15] Heisenberg did not derive this uncertainty relation from the Dirac–Jordan transformation theory, as he had done in the case of the position-momentum uncertainty relation. Unlike the observable, position, time-of-occurrence in quantum mechanics functions as a parameter which is not representable by a self-adjoint operator. For this reason the derivation and interpretation of the time-energy uncertainty relation is highly controversial. The standard derivation, from the properties of wave-packets, is due to Bohr.[16]

8.3
Einstein's delayed-choice experiment

Reflecting further on the photon-box experiment, Einstein pointed out that after the release of the photon from the box we may either re-weigh the box or open the box and compare the reading of the internal clock with the standard time scale, but not both. Hence we may either determine the energy of the liberated photon or the time of its release from the box, but not both; and so we may predict either the time at which the photon will arrive at a certain location or its energy at that location, but not both.[17] Moreover, we may do either of these things without disturbing the photon, since these determinations can be made *after* the photon has left the box.[18]

Einstein made this point in an article written jointly with R.C. Tolman and B. Podolsky.[19] Bohr was informed about it in a letter from Ehrenfest.[20] The philosophical significance of Einstein's point – which was not made clear either in his article or in Bohr's account – is that if we can determine either the energy or the time-of-occurrence of the process concerned *without disturbing the photon*, then these determinations are not mutually interfering; hence that process must really be characterised by exactly the energy and the time-of-occurrence that we could determine without disturbing it, even though these two attributes cannot both be observed. If we can determine either the exact energy or the exact time-of-occurrence of a process without disturbing it, then that process must *have* an exact energy and time-of-occurrence; there must be a definite state of physical reality which cannot be captured by quantum mechanics. The implication is, as Bohr puts it, that 'it might thus appear as if this formalism did not offer the means of an adequate description'.[21] This line of thought is, as Jammer has noted, the germ of the argument which Einstein developed in 1935, i.e. the EPR argument.[22]

As Bohr points out, a similar possibility arises in the case of the double-slit experiment. *After* the object has passed through diaphragm D_1 we may either re-measure the momentum of D_1 and thereby determine the momentum of the object, or we may establish a rigid connection between D_1 and the apparatus defining the spatial reference frame, thereby determining the position of the object at D_1 (provided D_1 is sufficiently massive). Hence *after* the object has passed through diaphragm D_1 we may either determine through which slit of D_2 the object passed, in which case no interference effects can be observed on the screen – or we may observe interference effects, foregoing the opportunity of determining through which slit the objects passed (see Figure 6). As Bohr observes, it makes no difference whether our plans for handling the apparatus are fixed in advance of the experiment or postponed until the object has passed through the diaphragm.[23] This imaginary experiment indicates that the observable behaviour of the object does not depend solely on what we do to the object; it may depend also on what we do to the experimental apparatus after the relevant object-instrument interaction has taken place. In this experiment the two possible, though mutually exclusive, measurements complete two distinct, well-defined phenomena. It was considerations of this sort that convinced Bohr that certain physical properties of microphysical objects are relative. In his discussion of delayed-choice experiments, J.A. Wheeler even maintains that in the double-slit experiment, the choice of what to measure may be delayed until the object has passed through the *second*, double-slit diaphragm.[24] W.C. Wickes and co-workers have undertaken to carry out a real delayed-choice experiment which is based on Wheeler's analysis of the double-slit experiment.[25]

Bohr's response to Einstein's point is clear. The box experiment as envisaged by Einstein is not a *single* experiment, but two different, mutually exclusive experiments, involving different experimental arrangements, a weighing device on the one hand, and a clock-work shutter on the other hand. For Bohr the box experiment illustrates once again the fact that any well-defined application of the quantum-mechanical formalism requires the specification of the whole experimental arrangement.[26] In an unusually clear discussion of this sort of experiment he says:

> ...no well-defined use of the concept of "state" can be made as referring to the object separate from the body with which it has been in contact, until the external conditions involved in the definition of this concept are unambiguously fixed by a further suitable control of the auxiliary body.[27]

Bohr's point is that the two mutually exclusive choices open to us refer, not as Einstein supposes, to one and the same state of *physical reality*, but to two different *phenomena* in Bohr's sense of that term.

8.4
The EPR experiment

In 1935 Einstein renewed his criticism, not against the epistemic interpretation of kinematic-dynamic complementarity, but against the ontic interpretation. In a famous paper written in collaboration with Podolsky and Rosen, he tried to show that the quantum-mechanical description of physical reality is in a certain sense incomplete.[28] The EPR argument, as it is now called, constitutes one of the most penetrating criticisms of the ontic interpretation of kinematic-dynamic complementarity ever given.

The physical basis of the argument is the following imaginary experiment. Two objects S_1 and S_2 are made to interact at a certain time t, after which they cease to interact, and then separate. If the states S_1 and S_2 before their interaction at time t were known, then the state of the combined system $S = S_1 + S_2$ – but not the state of the individual objects – could be calculated for any time after t.

The state of S may be expressed in either one of two ways:

(a) $\Psi(x_1, x_2) = \sum\limits_{m=1}^{\infty} \psi_m(x_2) v_m(x_1)$

where x_1 and x_2 are the variables used to describe S_1 and S_2 respectively, and where $v_m(x_1)$ are the eigenfunctions of an observable A of S_1 with eigenvalues a_m;

(b) $\Psi(x_1, x_2) = \sum\limits_{n=1}^{\infty} \phi_n(x_2) v_n(x_1)$

where $v_n(x_1)$ are the eigenfunctions of a physical quantity $B \neq A$ of S_1, with eigenvalues b_n.[29]

Now if on the one hand the quantity A of S_1 were measured and the value a_k found, then the state of S_1 after the measurement would be $v_k(x_1)$. It could then be inferred that the state of S_2 would be ψ_k, with corresponding value c_k. If on the other hand the quantity B of S_1 were measured and the value b_r found, then the state of S_1 would be $v_r(x_1)$, from which it could be inferred that the corresponding state and value for S_2 would be $\phi_r(x_2)$ and d_r respectively.

In the example which Einstein, Podolsky and Rosen consider, the two wave functions ψ_k and ϕ_r are taken to be eigenfunctions of the two non-commuting operators for momentum and position.[30] Thus a measurement of the momentum of S_1 would enable one to predict with certainty the momentum of S_2 without in any way disturbing the physical state of S_2; and a measurement of the position of S_1 would enable one to predict with certainty the position of S_2 without disturbing it. It may reasonably be assumed that neither the measurement on S_1 nor the calculation of the value of the corresponding physical quantity of S_2 would disturb the physical state of S_2 in any way, since S_1 and S_2 may be arbitrarily far apart when the measurement on S_1 is performed. The two measurements, it is also assumed, cannot both be performed at the same time.

A good example of the sort of situation which Einstein, Podolsky and Rosen have in mind is a system in the singlet state (i.e. total spin zero), composed of two spin-$\frac{1}{2}$ particles. If two such particles A and B have ceased to interact, and the x-component of the spin of A were measured and found to be $+\frac{1}{2}\hbar$, then, since the total spin is zero, it could be inferred that the x-component spin of B would be $-\frac{1}{2}\hbar$.[31]

An illustration of the EPR situation which is practically possible is provided by experiments on the polarisation of photons. Two photons, A and B, produced by the annihilation of an electron and positron at rest, separate along opposite directions. Each photon has both a circular, and a linear, polarisation. The state of the composite object A + B can be expressed either as:

$$\Psi = 2^{-\frac{1}{2}}(C_r^A C_r^B - C_l^A C_l^B)$$

with $C_r^A C_r^B$ a state containing two photons circularly polarised to the right; or as:

$$\Phi = 2^{-\frac{1}{2}}(L_x^A L_y^B - L_y^A L_x^B).$$

If the circular polarisation of A is found on measurement to be C_r^A, then the state of B can be inferred to be C_r^B. If the linear polarisation of A is found to be L_x^A, then the state of B can be inferred to be L_y^B. At any one time one may measure either the circular, or the linear, polarisation of A, but not both; hence one may predict either the circular, or the linear, polarisation of B, but not both.[32]

8.5
The EPR argument

The EPR argument is presented in a rather confusing manner. I propose to reformulate it in a way that clarifies the intentions of the authors.

The argument has two parts. In the first part the authors explain what they mean by a 'complete theory' by laying down a necessary condition of completeness: *every element of the physical reality must have a counterpart in the physical theory.*[33] What they mean by 'counterpart' is that an element of physical reality should be represented in a state description within the theory. The first part of the argument can be formulated in the following way:

(a) If a physical theory is complete, then, if x is an element of physical reality, there is a state description within the theory which includes x. (The completeness condition.)

(b) There are elements x, y of physical reality that are not both included in any quantum-mechanical state description.

(c) Therefore quantum mechanics is not a complete physical theory.

The second part of the argument purports to demonstrate premiss (b) above. Here the authors explicate what they mean by 'physical reality' by specifying a sufficient condition of physical reality:

> If, without in any way disturbing a system, we can predict with certainty (i.e., with probability equal to unity) the value of a physical quantity, then there exists an element of physical reality corresponding to this physical quantity.[34]

The term 'predict' is very misleading here, since no reference to the future is required by the argument; so I shall substitute the vaguer, but less misleading, term 'determine'.

Referring now to the EPR experiment, the authors argue that since we can determine with certainty either the position or the momentum of S_2 at time t, on the basis of a measurement on S_1, it follows *via* the criterion of physical reality that the position and momentum which can be determined with certainty for time t must be simultaneous elements of physical reality.[35] As it stands, this reasoning is fallacious. The truth of a disjunction does not entail the truth of the corresponding conjunction. From the fact that we can determine with certainty either the exact position or the exact momentum of S_2 at time t it does not follow by way of the reality criterion that S_2 has an exact position and an exact momentum at t. This argument, however, is not quite what Einstein had in mind. What he intended can be put as follows. Whether we determine at time t the position or the momentum of S_2, the physical state of S_2 at t remains the same, since neither a measurement on the distant S_1 nor the determination concerning S_2 can have any effect on the physical state of S_2. Hence, if we determine the position of S_2 at t, then S_2 must have at t whatever value of the momentum we would have determined had we so chosen; and conversely, if we determine the momentum of S_2 at t, then S_2 must have at t whatever value of the position we would have determined had we chosen to determine the position. From what he says elsewhere, it is clear that this is the argument that Einstein had in mind.[36] The second argument can be formulated as follows:

(1) We can determine either the exact position or the exact momentum of S_2 at t, but not both.

(2) The real physical state of S_2 is the same, whether we determine the exact position or the exact momentum of S_2.

(3) Therefore there is at t a single real physical state of S_2 in which position and momentum both have exact values.

(4) Operators representing position and momentum are non-commuting.

(5) Therefore, there exists a single physical state in which two physical

quantities represented by non-commuting operators have exact simultaneous values.

(6) The physical state of an object at any time is completely described by a single state vector.

(7) Different non-commuting operators have no state vectors in common.

(8) Therefore a physical state in which physical quantities represented by non-commuting operators have exact simultaneous values is not describable in terms of a single state vector.

(9) But such a physical state exists, viz. the one referred to in premiss (5).

(10) Therefore there are elements of physical reality, x, y, which are not included in any quantum-mechanical state description. (Premiss (*b*) of the previous argument.)

Einstein rightly treats premisses (1), (4), (6), (7) and (8) as forming part of the orthodox interpretation of quantum mechanics; only premisses (2), (3), (5), (9) and (10) are possibly controversial from the orthodox point of view.

It is a corollary of the EPR argument that it is possible to assign to the same physical reality two different state vectors which are eigenfunctions of non-commuting operators; this corollary is clearly stated, indeed it is emphasised in the EPR paper.[37] Contrary to the orthodox view, then, it cannot be the case that such vectors always refer to different physical states. If it were the case, we should have to say that either the measurement on S_1 or the determination concerning S_2 has some physical effect on S_2. Einstein brought out the significance of this corollary more clearly in later writings; indeed, it is the point that he stresses.[38] The point is that the orthodox interpretation of quantum mechanics can be upheld only if the measurement on S_1 or the determination concerning S_2 somehow affects the physical state of S_2. But how can it? Such an effect would require a curious action at a distance. This curious situation is known as the 'EPR paradox': if the physical state of an object is completely described by the state vector, and if state vectors belonging to two non-commuting operators describe different physical states of the object, then the physical state of an object can be changed by making a measurement on some other, distant object. The completeness thesis entails that the real states of spatially separated objects are not independent of each other.[39]

A year after publication of the EPR paper, Einstein brought his statistical interpretation to bear on the EPR paradox. The state vector, he argued, does not refer to the physical state of a single object, but to a statistical ensemble of objects. In assigning different state vectors to S_2 on the basis of different measurements on S_I (were this possible), we would not be changing the physical state of S_2; rather, we would be assigning S_2 to different statistical ensembles. These quantum-mechanical ensembles, however, cannot be ultimate, if an object in the same physical state can be assigned to different ensembles. It was Einstein's hope that a future theory would be able to provide a more complete description of physical reality than quantum mechanics can provide.[40]

8.6
Bohr's response to the EPR argument

Bohr's initial response to the EPR paper was published in a preliminary note to *Nature*. He did not accept the conclusion of the EPR argument, that quantum mechanics is not a complete theory. He objected that one of the premises of the argument, the criterion of physical reality, contains 'an essential ambiguity' when applied to the problems of quantum mechanics. The measurement on S_I, he admits, has no physical effect on S_2; it has nevertheless an essential influence on the necessary conditions for the definition of the physical properties concerned:

> Since these conditions must be considered as an inherent element of any phenomenon to which the term 'physical reality' can be unambiguously applied, the conclusion of the above-mentioned authors would not appear to be justified.[41]

In his definitive rejoinder to Einstein, Bohr again states that the flaw in the EPR argument is the ambiguity of the EPR criterion of physical reality.[42] What is this ambiguity? In order to explain what he means, Bohr analyses the double-slit experiment.

He argues that, depending on whether the diaphragm is rigidly or loosely attached to the rest of the apparatus, the double-slit arrangement allows one to measure either the position or the momentum of an object that passes through the diaphragm, but not both.[43] He argues that it is not just that the conditions required for measuring exact position and exact momentum are mutually exclusive, but rather that the conditions for the well-defined applicability of these concepts are mutually exclusive. We ought not to say that exact knowledge of the one property entails ignorance of the other, for we can meaningfully be said to be ignorant only of what is there to be known:

if one property is known, then talk of the other property, even as 'necessarily unknown', is not well defined. Thus Bohr writes:

> . . . in the phenomena concerned we are not dealing with an incomplete description characterized by the arbitrary picking out of different elements of physical reality at the cost of sacrificing [*sic*] other such elements, but with a rational discrimination between essentially different experimental arrangements and procedures which are suited either for an unambiguous use of the idea of space location, or for a legitimate application of the conservation theorem of momentum . . . Indeed we have in each experimental arrangement suited for the study of proper quantum phenomena not merely to do with an ignorance of the value of certain physical quantities, but with the impossibility of defining these quantities in an unambiguous way.[44]

In later writings Bohr illustrates this point by employing the single-slit experiment as a model of the EPR experiment. Imagine a heavy diaphragm S_1 with a single slit through which passes particle S_2. Now since the position of co-ordinate y_1 of S_1 commutes with the momentum component p_y^2 of S_2, and y_2 with p_y^1, the difference $y_1 - y_2$ commutes with the sum $p_y^1 + p_y^2$. Hence the quantities $y_1 - y_2$ and $p_y^1 - p_y^2$ are each calculable. If the momenta of S_1 and S_2 were exactly known before S_2 passes through S_1, then after the passage of S_2 we could *either* measure again the momentum of S_1, in which case the momentum of S_2 could be inferred by applying the law of the conservation of momentum; *or* we could measure the position of S_1; in which case we could calculate the position of S_2 at the moment of contact with S_1 and at any future time, provided S_1 were sufficiently heavy that the uncertainty in its momentum would not appreciably affect its velocity. But we could not do both.[45]

If on the one hand we were to re-measure the momentum of S_1, its exact position when S_2 passed through the slit could not be determined. If on the other hand we were to measure the position of S_1 after the passage of S_2, by connecting S_1 rigidly to the laboratory frame, then the energy exchanged between S_2 and S_1 could not be determined. Bohr stresses the point that either, but not both, of these two procedures can be carried out *after* S_2 has passed through S_1 and consequently without interfering with S_2 in any way.[46]

This is exactly the situation that troubled Einstein as early as 1931. Einstein was inclined to argue that since either the exact momentum of S_2 or the exact position of S_2 can be determined in this case without physically disturbing it in any way, S_2 must be said to *have simultaneously* an exact position and an exact momentum, contrary to Bohr's interpretation of the uncertainty relations. As Bohr remarks, it might seem, as it does to Einstein, that the two possible determinations (or predictions) must refer to the same physical state of S_2, and ought therefore to be included in 'an exhaustive account of physical reality': it might seem that we are 'able to assign to one and the same state of the object two well-defined physical attributes in a way incompatible with the uncertainty relations.'[47] Bohr maintains, however, that this is not the case; for we are concerned here with '*a discrimination between different experimental procedures which allow of the unambiguous use of complementary classical concepts*'.[48]

Although a measurement on S_1, Bohr admits, could not physically disturb S_2, nevertheless the experimental arrangement required for the measurement determines both for S_1 and for S_2 the necessary conditions for the meaningful ascribability of the physical property concerned. For the properties of exact position and exact momentum, moreover, these conditions mutually exclude each other. If the momentum of S_1 were measured, then the necessary conditions for the meaningful ascribability of an exact position to S_1 and S_2 (or even of meaningful talk of such a property) would not be satisfied. *Mutatis mutandis* the same holds for a measurement of the position of S_1. Hence, Bohr holds, Einstein is mistaken in assuming that S_2 must have both an exact, though unknown, simultaneous position and momentum. An object cannot meaningfully be said to have certain properties in the absence of the conditions which make such talk meaningful. This is what Bohr means when he writes:

> Of course there is in a case like that just considered no question of a mechanical disturbance of the system under investigation during the last critical stage of the measuring procedure. But even at this stage there is essentially the question of *an influence on the very conditions which define the possible types of predictions regarding the future behaviour of the system*. Since these conditions constitute an inherent element of the description of any phenomenon to which the term "physical reality" can properly be attached, we see that the argumentation of the mentioned authors does not justify their conclusion that quantum-mechanical description is essentially incomplete.[49]

In the phrase 'during the last critical stage of the measuring procedure' Bohr is referring to the decision in a delayed-choice experiment about whether to fix the apparatus for a position measurement or a momentum measurement.

The notion of the well-defined physical state of an object is, Bohr maintains, not well defined unless the necessary conditions for the meaningful applicability of the properties defining the state are satisfied. We properly ought not to say, as Einstein does, that the determination of either the position or the momentum of S_2 (whichever the case may be) must refer to the same well-defined physical state, for in this situation the notion of the 'same well-defined physical state' is not well defined. Thus Bohr disputes premiss (2), Section 8.5 in the second part of the EPR argument.

In his 'Autobiographical Notes' Einstein writes:

> Now it seems to me that one may speak of the real factual situation of the partial system S_2 . . . the real factual situation of S_2 is independent of what is done with the system S_1, which is spatially separated from the former.[50]

Bohr, however, holds that there is an ambiguity in saying that the real factual situation of S_2 is independent of what is done with S_1, for in one sense this is correct, viz. in the sense that the physical operations which we actually perform on S_1 when making a measurement have no causal effect on the physical state of S_2; but in another sense it is false, viz. in the sense that the real state of S_2 is not independent of the experimental arrangement that we choose to set up, for the set-up may determine the conditions which the 'factual situation' of S_2 presupposes.

Although S_2 is not physically disturbed in the EPR experiment, it would be wrong to think that the notion of the measurement disturbance plays no significant role in the argument, for S_1 is inevitably disturbed by whichever measurement is performed upon it, and it is this disturbance, among other factors, which prevents us from assigning to S_2, without disturbing it, a well-defined position and momentum at the same time.

Einstein's incisive criticism led Bohr to emphasise points which he had not stressed before. He emphasised that the quantum-mechanical formalism cannot be given a well-defined application independently of the conditions, determined by the *whole* experimental arrangement, which are necessary for the meaningful applicability of the physical concepts: the application of quantum mechanics is perfectly well defined only when it refers to a *phenomenon*, to the object *cum* set-up defining the necessary conditions for the applicability of the physical concepts employed. In the EPR experiment determinations of the position or the momentum of S_2 refer to different, indeed complementary, phenomena (though of course to the same object).[51] He preferred also to say that the uncertainty relations express an unavoidable limitation on the joint *applicability* of two well-defined concepts, rather than a limitation on the joint *measurability* of two well-defined attributes.[52] This is what he has in mind when he says:

> ... the whole situation in atomic physics deprives of all meaning such inherent attributes as the idealizations of classical physics would ascribe to the object.[53]

What he means by 'inherent' here is 'absolute' in the sense of being intrinsic to the object irrespective of any external conditions and its relation to them. He may in fact have held this view from the outset, his defence against Einstein's criticism having helped simply to clarify and re-inforce it. In the late twenties we find him saying that 'an independent reality in the ordinary physical sense' cannot be attributed to the object; by which he may have meant that the properties of the object cannot be said to be absolute or wholly intrinsic in the sense of existing independently of certain conditions (frame of reference etc.) which the meanings of the concepts presuppose.[54] This is what he had in mind when he said that the uncertainty relations do not express a limit to the *extent* of the information obtainable through observation but 'a limit to the *meaning* which we may attribute to such information'.[55]

The main new insight which he gained through Einstein's criticism was a clear awareness of the importance of distinguishing between the notion of the interaction between the object and the instrument and the notion of the experimental arrangement's determining the conditions for the meaningful application of the physical concepts.

8.7
Einstein's response to Bohr's defence

How did Einstein understand Bohr's response to his criticism? In the EPR paper it is taken for granted that, in the light of the argument, there is no alternative to abandoning the orthodox view that quantum mechanics provides a complete description of the state of an individual object. The authors note, merely as an afterthought, that one might save the orthodox interpretation by abandoning instead the view that the physical reality of the position and momentum of S_2 is independent of the process of measurement carried out on S_1; but they do not consider this to be a viable alternative.[56]

After Bohr's rejoinder, however, Einstein treats the latter view seriously as a genuine, though for him untenable, position.[57]

Don Howard argues that the EPR paper does not accurately convey Einstein's intentions. In a letter to Schrödinger of 19 June 1935, which Howard cites, Einstein remarks that the EPR paper was written by Podolsky after many discussions, and adds: 'What I really intended has not come across very well; on the contrary the main point was, so to speak, buried by erudition'.[58] It is clear from this letter that the main point which Einstein thought had been obscured in the EPR paper was what he saw as the crucial role of the *separation principle*, which is to the effect that the real state of an object cannot depend upon the kind of measurement we make on another, spatially separated object. He proceeds to present the argument based on this principle. The argument, as I understand it, is this:[59]

(1) A description of the real state of an object is complete only if it is unique – there can be only *one* complete description of an object.

(2) Different state vectors describe different real states of an object.

(3) In an EPR-type situation we may assign different state vectors to one and the same real state of an object.

(4) Therefore, neither of these state vectors provides a complete description of the state of an object.

The conclusion (4), however, can be denied if in premiss (3) it is denied that the real state to which the two different state vectors are assigned is the *same* state of the object. In order to deny this, however, one is required to reject the separation principle; for, if in this case the two different state vectors are assigned to different real states of the object, then the real state of the object is determined either by the assignment of the state vector to it or by the measurement on the other, spatially separated object on which the assignment is based; in which case the separation principle is violated. Einstein's main point, then, is that the separation principle must be included as an explicit premiss in the argument.

The above argument, as Howard points out, is different from the argument as it is expounded in the EPR paper; but it is not, as Howard maintains, 'strikingly different'.[60] Admittedly the above argument is not explicitly spelled out in the EPR paper, but its embryo is there. Premiss (3) is what I have called a corollary to the main argument, and it is clearly stated, indeed emphasised, on page 779 of the paper; and the separation principle is put to use on the last page:

This makes the reality of P and Q depend upon the process of measurement carried out on the first system, which does not disturb the second system in any way. No reasonable definition of reality could be expected to permit this.[61]

If the reference to the separation principle here was an afterthought, it was not, judging by his letter to Schrödinger, Einstein's afterthought, but Podolsky's.

The impression might be got from Howard's paper that Einstein disowned the EPR argument; such an impression, however, would be quite mistaken. Einstein's dissatisfaction with the EPR paper was simply that it did not do justice to all the points he wished to make. He summarised the EPR argument in a letter to Popper of 11 September 1935. He first presents the argument in much the same way as he does in his letter to Schrödinger, i.e. couched in terms of different wave functions; and then he goes on to recount the argument as it is given in the EPR paper, i.e. couched in terms of different values of observables.[62] There can be little doubt, however, that it was the later argument, as distinct from the main argument of the EPR paper, which he thought was the more important. By 'EPR argument' I shall mean both arguments.

Howard claims also that implicit in Einstein's thought at the time when the EPR paper was produced was not merely the separation principle but also the *principle of separability*, as Howard calls it, which is to the effect that spatially separate objects have their own separate states. In Howard's view the 'main point' referred to in Einstein's letter to Schrödinger is 'the need for an explicit separability assumption'.[63] Here again I part company with Howard. In later writings Einstein employs an assumption that might be called the *principle of independent existence*, according to which the real states of spatially separated objects are independent of each other. Einstein came to recognise a distinction between the separation principle and the principle of independent existence some time *after* he had studied Bohr's reply to the EPR paper. The first reasonably clear sign that Einstein recognised a distinction between these two assumptions appears in his 'Autobiographical Notes' which, as Howard points out, was written probably between March 1946 and March 1947.[64] What brought the distinction home to Einstein was the thought that Bohr did not wish to reject the separation principle. Einstein treats the separation principle as being equivalent to a principle of local action which states that occurrences taking place where one object is situated can have no instantaneous causal effects on another, spatially separated object. Now Bohr clearly did not wish to deny the principle of local action; so it seemed to Einstein, I suggest, that what Bohr wished to deny was the principle of independent existence.

In his 'Replies to Criticisms' Einstein gives a clear account of what he sees as Bohr's reply to the EPR argument. Bohr holds that if the partial objects S_1 and S_2 form a total system with a wave function of its own, there is no reason why S_1 and S_2, when viewed separately, should be ascribed 'any mutually independent existence (state of reality)', not even when they are spatially separated from each other. Einstein's rejoinder is that we may choose to hold either that the description provided by the state vector is complete or that the real states of spatially separated objects are independent of each other, but not both. Einstein chose the latter, Bohr the former.[65]

Was Einstein right in his reading of Bohr's reply? Did Bohr in fact wish to reject the principle of independent existence? Whether or not he did, he did wish to reject the separation principle, though not as Einstein understood it, as being tantamount to a principle of local action. Bohr accepted the principle, thus construed; but he rejected it, construed in another, Bohrian way, to the effect that what the physical state of an object can meaningfully be said to be may depend upon conditions determined by the experimental arrangement which we set up for measurements on another, spatially separated object; in this sense the physical reality of one object may depend upon what we do to or at another, spatially separated object. Bohr, however, does not think that this sort of dependence involves any causal interaction between the objects in question, any sort of action at a distance; rather, what we do at one object may create a physical context which determines what we can meaningfully say about some other, spatially separated object. Einstein, understandably, was never able to grasp Bohr's point here, and hence took him to be denying the principle of independent existence. It is interesting to note, however, that of all the orthodox positions with which Einstein was acquainted, Bohr's seemed to him 'to come nearest to doing justice to the problem'.[66] The question whether Bohr wished also to reject the principle of independent existence is one which I shall discuss in the following chapter.

8.8
A preliminary summing-up

Einstein's argument is very powerful, much more powerful than is generally recognised. If different state vectors describe different physical states of an object, then in the EPR experiment the physical state of an object, S_2 say, differs depending upon whether we assign one or another state vector to it; it seems that we may alter the physical state of S_2 without acting upon it in any way – a paradoxical state of affairs indeed. The paradox is dissolved if we hold that the state vector provides only a partial, and not a complete, description of the state of an object. Again, if we may determine with certainty either the exact position or the exact momentum of S_2 at time t without disturbing S_2, then it is plausible to hold that S_2 has at t both the exact position and the exact momentum either of which we could determine just as we pleased. Now the orthodox view is that we cannot say that an object must have both the position and the momentum either of which we could measure if we so chose, because measured values are created by the process of measurement. In the EPR situation, however, the orthodox view cannot be saved by appeal to the creation-by-measurement doctrine, for, as Einstein shows, in this case the creation by measurement would involve a counterintuitive action at a distance. The creation-by-measurement doctrine is puzzling enough, but in the EPR situation it becomes barely intelligible. Bohr clearly saw this; hence he rejected the doctrine.

Bohr has a ready reply to Einstein's argument: we cannot meaningfully ascribe a physical property to an object unless the preconditions for the meaningful use of the predicate in question are satisfied, and these preconditions are the presence of an appropriate experimental arrangement that is capable of being used to measure the property in question (the weak meaning condition). Some time after his reply to the EPR paper was published, Bohr, I suspect, came to see a difficulty with his reply, which is that in the EPR experiment it seems that the preconditions for the meaningful use of the physical predicates *can* be satisfied: we can set up an instrument for measuring the position of S_1, and an instrument for measuring the momentum of S_2. In view of this apparent difficulty, Bohr added a further precondition for the meaningful ascription of a physical property, which is that an actual measurement of the property in question should have been performed (the strong meaning condition). The mere presence of an appropriate measuring instrument, though necessary, is not sufficient in itself to constitute a well-defined phenomenon in Bohr's technical sense. As we noted in the previous chapter, Bohr made it clear in his Warsaw lecture that a phenomenon is well defined only if a measurement has actually been performed, and the notion of the state of an object is well defined only when applied in a context which includes the performance of a measurement.[67]

Bohr, however, had no real need to introduce the strong meaning condition, for the putative difficulty is spurious: we can set up an instrument to measure the position of S_1 and another instrument to measure the momentum of S_2, but these instruments cannot be *used* to make measurements in such a way that the weak meaning condition is satisfied. Say we measure the momentum of S_2 at time t; now it might seem that if we also measure the position of S_1 at t, then on the basis of this measurement we can meaningfully ascribe a position to S_2 at t. This, however, can not be done; for, if at time t we measured the position of S_1 and the momentum of S_2, the position measurement would involve a transfer of momentum from S_1 to the body defining the common spatio-temporal reference frame; and this transfer would interfere with the measurement of the momentum of S_2.[68] In the single-slit version of the EPR experiment, for example, the instrument which is set up to measure the position of the object S_2 is in fact not capable of being used to make such a measurement if a simultaneous measurement of the momentum of the diaphragm S_1 is to be made. The reason for this is that a measurement of the position of S_2 would involve an exchange of energy and momentum with the system defining the spatio-temporal reference frame, and so the measurement of the momentum of S_1 would no longer relate to the same frame. The strong meaning condition, therefore, is superfluous; the weak meaning condition suffices.

If my suggestion concerning Bohr's motive for introducing the strong meaning condition is correct, it may seem that his introduction of this condition was purely *ad hoc*, designed solely to save the indefinability thesis from refutation in the light of the EPR argument. The criticism, however, would not be entirely just, for, as we have seen, in the late twenties Bohr was inclined to think that ascriptions of values to observables which were not based upon measurement operations were mere abstractions, devoid of any realistic content. By adding in the late thirties that the performance of a measurement is required for a well-defined phenomenon, Bohr can be seen merely to be making explicit a point which was implicit in his earlier views. Bohr had genuine, *non-ad hoc* reasons for proposing the strong meaning condition, but since they are primarily philosophical, I shall postpone spelling them out until Chapter 10, where his philosophy will be discussed in detail. A justification is required also for his introduction of the weak meaning condition; his grounds for this condition are similar to his grounds for the strong condition. Until these grounds have been presented, a verdict on the Bohr–Einstein debate would be premature. Besides, there are further developments relating to the physics which we shall have to examine before a verdict can be given; this task will be undertaken in the following chapter. What I can say at this stage is that Einstein's argument is very powerful; and in his rejoinder to it, Bohr merely lays down the weak and strong meaning conditions, without providing adequate justification for them. Einstein conceded that he had not shown that quantum mechanics is incomplete, but rather that it can be held to be complete only at the high, and for him exorbitant, cost of abandoning either the separation principle or the principle of independent existence.

9

The sequel to
the Bohr–Einstein debate

The EPR paper and Bohr's reply to it brought to an end the public debate between Einstein and Bohr. In the aftermath, Bohr's construal of quantum mechanics established itself as the basis of the orthodox interpretation – the 'Copenhagen interpretation' as it came to be called. The vast majority of physicists regarded Einstein's criticism as reactionary, and dismissed it with disdain. One of the main causes of this disdainful attitude was the general opinion that Einstein's criticism committed him to the existence of hidden states, and the existence of these, it was thought, had been conclusively ruled out by the work of von Neumann and others. What light does the subsequent work on the question of hidden states, and in particular the work of Bell, throw on the Bohr–Einstein dispute?

9.1
Completeness and hidden states

There are three distinct questions concerning hidden states. First, might there be such states, unbeknown to quantum mechanics? Second, can quantum mechanics, as it stands, take account of such states? Third, could some theory containing descriptions of such states reduplicate the predictions of quantum mechanics? Concerning the first of these questions, Einstein's view was that it is very likely that such states exist, and since quantum mechanics takes no account of them, it is incomplete and best regarded as a statistical theory analogous in some respects to classical statistical mechanics. On this view the state vector describes the state of a statistical ensemble of objects rather than the state of an individual object; it represents a sort of average of the hidden states; and if quantum mechanics can take no account of such states, then some other theory might well be able to do so. The sort of hidden states which Einstein was interested in were states in which every observable of an object has an exact value at any instant; thus the sort of hidden-states theory he contemplated incorporated the intrinsic-properties theory. The EPR argument was intended to show that an affirmative answer to the first question must be given.

Concerning the second question, various arguments have been proposed which are designed to show that quantum mechanics cannot be supplemented with hidden states, in the sense that state vectors can be represented as averages over hidden states defined on a phase space, states in which every observable has an exact value. Such states would generate dispersion-free probability measures, i.e. measures such that for any state W, any observable A, and any value a, the probability that A will be found to have the value a in state W is 1 or 0. The question, then, whether such hidden states can be introduced into quantum mechanics is equivalent to the question whether the non–dispersion-free states of quantum mechanics W are resolvable into weighted sums, or mixtures, of pure and dispersion-free states W', W'', . . ., such that, for every observable A and value a,

$$\text{prob}_W(A = a) = c_1 \text{prob}_{W'}(A = a) + c_2 \text{prob}_{W''}(A = a) + \ldots$$

The definitive answer to this question is provided by a corollary to Gleason's theorem to the effect that, for every Hilbert space of three or more dimensions, every measure w (i.e. countably additive map of the subspaces of a Hilbert space into the non-negative real numbers) is expressible in the form $w(H) = \text{Tr}(WP)$, with H a subspace, W a statistical operator, and P a projection operator whose range is H. The corollary is that there are no dispersion-free measures on such a Hilbert space. So, given the Hilbert space formulation of the theory, quantum mechanics can contain no dispersion-free probability measures, and hence no pure and dispersion-free states.[1] (Note that the corollary is that there are no dispersion-free states with respect to *every* observable, but not that there are no such states with respect to *some* observables.) The work of Gleason, then, establishes that hidden states cannot be introduced into quantum mechanics, and that any theory which incorporates such states must have a different mathematical structure from that of quantum mechanics. This result would not have troubled Einstein, however, since he held all along that quantum mechanics is a provisional theory and would be superseded by a unified field theory which took account of the quantum of action. Gleason's work, then, leaves the first question – the question with which Einstein was most concerned – untouched; it leaves open the question of the completeness of quantum mechanics, and hence also the question whether the intrinsic-properties theory is true.

The third question was answered in the affirmative in 1952, when Bohm invented a hidden-states theory which re-duplicated the predictions of quantum mechanics, a result that would have been of no surprise to Einstein.[2] Bohm's theory, as we might expect, differs from quantum mechanics. Thus the results concerning hidden states that had been established by the early sixties provided no adequate grounds for the prevailing disdainful attitude towards Einstein's critique of the orthodox interpretation of quantum mechanics. A completely new dimension, however, was introduced into the question of hidden states by Bell in 1964.

9.2
Completeness and non-locality

As we have seen, Einstein maintained that quantum mechanics could be regarded as a complete theory only at the cost of abandoning a locality assumption. He had assumed that a hidden-states theory of the intrinsic-properties type did not require the abandonment of the separation principle. This assumption, however, was called into question by Bell. Analysing the quantum-mechanical statistics of Bohm's version of the EPR experiment, Bell argued that if hidden-states theories are to reproduce the predictions of quantum mechanics, then contrary to Einstein's assumption, they must violate a principle of local action, a locality condition to the effect that the measured value of an observable of an object is not affected by the sort of measurement which we make on another, spatially separated object with which it is paired (Bell's locality condition).[3] The following account of Bell's argument is based upon Wigner's simpler reformulation of it.[4]

Consider a pair of spin-$\frac{1}{2}$ particles in the singlet spin state. After an initial interaction at a common source each particle of a pair flies off in a direction opposite to that of the other. According to quantum mechanics the components of the spin vectors of both particles in any one direction are negatively correlated. If the spin-component of the particle that travels to the left (the L-particle) is $+\hbar/2$ along axis A, then the spin-component of the other particle (the R-particle) is $-\hbar/2$ along the same axis, and vice versa (the spin-component has only two values). There are, however, no such invariable negative, or positive, correlations of the spin-components along different axes: if the spin-component of the L-particle is $+\hbar/2$ along axis A, then the spin-component of the R-particle along a different axis, B or C say, may be either $+\hbar/2$ or $-\hbar/2$, the probability of either result being specified by the theory. Thus the probability that the spin-component of an L-particle will be found to be $+\hbar/2$ along axis A and the spin-component of the paired R-particle will be found to be $+\hbar/2$ along a different axis B is:

$$\mathrm{prob}(A+,B+)=\tfrac{1}{2}\sin^2(\tfrac{1}{2}\theta_{AB}),$$

where θ_{AB} is the angle between coplanar axes A, B. Similarly, for the axes A, C and B, C the probabilities are respectively:

$$\mathrm{prob}(A+,C+)=\tfrac{1}{2}\sin^2(\tfrac{1}{2}\theta_{AC}); \; \mathrm{prob}(B+,C+)=\tfrac{1}{2}\sin^2(\tfrac{1}{2}\theta_{BC}).$$

Where $\theta_{AB}=\theta_{BC}=60°$, and $\theta_{AC}=120°$, the probabilities are: $\mathrm{prob}(A+,B+)=\mathrm{prob}(B+,C+)=\tfrac{1}{8}$, and $\mathrm{prob}(A+,C+)=\tfrac{3}{8}$. Thus we have the inequality:

$$\mathrm{prob}(A+,C+) \geqslant \mathrm{prob}(A+,B+)+\mathrm{prob}(B+,C+) \tag{1}$$

Wigner, following Bell, argues that a hidden-variables theory can yield the inequality (1) only if it violates Bell's locality condition to the effect that the results of measurement on the L-particles are independent of the direction in which measurements are made on the R-particles, and vice-versa. Wigner's argument applies also to the intrinsic-properties theory, which is simply a hidden-variables theory, the hidden variables in this case being the pre-existing values of the observables which are hidden to quantum mechanics. Now if the intrinsic-properties theory is true, then, given the negative correlations between the spin-components of members of a pair of particles along any one direction, the frequencies of the measurement results must be such that the number $N(A+,\ C+)$ of pairs of particles whose spin-components are found on measurement to be $+$ along the different axes A and C (i.e. along A for the L-particles, and along C for the R-particles) is the sum of the numbers of pairs of particles whose spin-components in the directions A, B, C are either $++-$ and $+--$ for the L-particles, and $--+$ and $-++$ for the R-particles. Thus,

$$N(A+,C+)=N(++-,--+)+N(+--,-++),$$

where '$N(++-,-++)$' means 'the number of pairs of which the spin-components of the L-particles along A, B, C are $++-$, and the spin-components of the R-particles along A, B, C are $-++$.' Similarly,

$$N(A+,B+)=N(+-+,-+-)+N(+--,-++),$$

and

$$N(B+,C+)=N(++-,--+)+N(-+-,+-+).$$

The numbers of pairs $N(A+,C+)$ and $N(A+,B+)$ share pairs of the type

$(+--,-++)$; and the numbers of pairs $N(A+, C+)$ and $N(B+, C+)$ share pairs of the type $(++-, \; --+)$. If we adopt Bell's locality condition, then it is reasonable to assume that the numbers $N(+--, \; -++)$ are the same in both $N(A+, C+)$ and $N(B+, C+)$, then we have the following inequality (the Bell–Wigner inequality):

$$N(A+, C+) \leqslant N(A+, B+) + N(B+, C+). \tag{2}$$

This inequality is a consequence of the following inequality:

$$N(++-,--+) + N(+--,-++) \leqslant N(+-+,-+-) +$$
$$N(+--,-++) + N(++-,--+) + N(-+-,+-+). \tag{3}$$

This is the relation between the statistical frequencies that we should expect on the basis of the intrinsic-properties theory. Yet these statistics are markedly different from those predicted by quantum mechanics, viz. formula (1) above.

It is clear from these results that the statistics entailed by the intrinsic-properties theory could be equivalent to those entailed by quantum mechanics only if one of two alternatives obtained. Either a) no pairs of the type $(+-+,-+-)$ were present when (A, B) measurements were made, and no pairs of the type $(-+-, \; +-+)$ were present when (B, C) measurements were made; but this condition would obtain, surely, only if occurrences of pairs of the types $(+-+, -+-)$ and $(-+-, +-+)$ were determined by the choice of directions in which measurements were made; or b) no pairs of the type $(+--, -++)$ were present when (A, B) measurements were made, and no pairs of the type $(++-, --+)$ were present when (B, C) measurements were made; but this condition, surely, would be satisfied only if the presence of pairs of the types $(+--, --+)$ and $(++-, \; --+)$ was determined by the choice of the directions selected for measurements, i.e. only if Bell's locality condition were violated.

The conclusion of the Bell–Wigner argument is that if a hidden-states theory is to reproduce the predictions of quantum mechanics, then it must violate a locality condition. The argument purports to rule out the possibility that a theory of the intrinsic-properties type should satisfy a locality condition and reproduce the well-confirmed predictions of quantum mechanics: such a theory must either violate a locality condition or be empirically inadequate. This outcome would have been unpalatable to Einstein, for he would have been loath to relinquish either a locality condition or the requirement that a hidden-states theory should be empirically adequate. Einstein wished to give affirmative answer both to the question 'Might there be hidden states?' and to the question 'Could some local hidden-states theory reduplicate the predictions of quantum mechanics?'. An affirmative answer to the former question and a negative answer to the latter would not have appealed to Einstein.

The conclusion of the Bell–Wigner argument, however, is not absolutely watertight, since the argument contains a tacit assumption which may be questioned, viz. the assumption that the frequency with which any value of an observable is obtained in measurements is the same as the pre-measurement distribution of that value (the 'matching condition'). There are at least two ways in which this tacit assumption might be false. First, the assumption presupposes the objective-values theory. But this theory might be false, either in general, as Georges Lochak argues that it is,[5] or in the special case of spin measurements. In either case, the measured values of the spin-components would not be pre-existing values, and hence the frequency with which any value was obtained in measurements would not be the same as the pre-existing distribution of that value. In this case a hidden-states theory would not be required to reproduce the predictions of quantum mechanics, and hence need not violate a locality condition. But since in this case the hidden states would not generate the correct probabilities for the results of measurements, they would be empirically otiose. Such a theory would be of metaphysical interest only – not a favourable consequence for Einstein. Second, as Arthur Fine points out, the assumption that the frequency with which any value is obtained in measurements is the same as the pre-measurement distribution of that value is not entailed by the objective-values theory: the frequencies of the measured values may differ from the frequencies of the pre-existing values, even on the assumption of the objective-values theory. The two frequencies in question would in fact differ if, owing to some restriction of which we are at present unaware, not every pair of particles produced by the source is available for measurement, or if paired particles are not perfectly synchronised once they leave the source. Fine maintains that this situation may well obtain, and he proposes models – 'prism models' – which depict it.[6] The advantage of Fine's hypothesis is that it retains the objective-values theory: the disadvantage is that it is implausible; since a mechanism which excludes some types of pairs would depend upon the directions we choose for measurement – an unlikely event.[7]

If the tacit assumption of the Bell–Wigner argument is false, as it may be, the argument is irrelevant to the Bohr–Einstein debate. If, however, the assumption is true, then the argument is more damaging to Einstein's position than it is to Bohr's; indeed, it leaves Bohr's position intact, since Bohr denied that a hidden-states theory of the intrinsic-properties type is a genuine possibility, and he held no strong views on the question of locality. The sequel to the debate, then, is more favourable to Bohr than to Einstein.

9.3

The scope of non-locality

What exactly is the scope of the non-locality which must, it seems, be a feature of hidden-states theories of the intrinsic-values/objective-values sort? Bell seems to have assumed that the locality condition is a problem only for hidden-states theories. Is this assumption correct? In order to answer this question we shall have to distinguish between different kinds of probability condition. Jon Jarrett argues that the locality condition which is generally employed in order to derive Bell-type inequalities is a conjunction of two logically independent conditions. The first of these, which Jarrett calls simply 'the locality condition', is to the effect that the result of a measurement on one of a pair of spin-$\frac{1}{2}$ particles is statistically independent of the setting of the instrument used to make a measurement on the other particle of the pair, or more precisely, the probability of a certain measurement result in any direction on the L-wing, given that no measurement is made on the R-wing, is equal to the probability of a certain measurement result on the L-wing in any direction, given that a measurement is made on the R-wing, and vice versa. In formal terms: for all states W, and all directions d^L, d^R relating to the L-particles and the R-particles respectively, and for all values $d^L_i = +, -, d^R_j = +, -,$

$$\text{prob}_W(d^L_i; o^R) = \text{prob}_W(d^L_i; d^R); \text{ and } \text{prob}_W(d^R_j; o^L) = \text{prob}_W(d^R_j; d^L),$$

where the expression '$\text{prob}_W(d^L_i; o^R)$' denotes the conditional probability that a measurement of the d spin-component of an L-particle in state W yields the value i, given that no measurement is made on the paired R-particle.[8] Following Jarrett, I shall call this condition 'the locality condition' (or 'Jarrett's locality condition').

The second condition which Jarrett distinguishes is to the effect that the results of measurements on both particles of a pair are statistically independent of each other: the probability of a certain measurement result on the L-wing is independent of the result of measurement on the R-wing. In formal terms: for all states W, all measurements in directions d^L, d^R, and all values $d^L_{i=+,\ -}$, $d^R_{j=+,\ -}$,

$$\text{prob}_W(d^L_i,\ d^R_j) = \text{prob}_W(d^L_i;\ d^R) \times \text{prob}_W(d^R_j;\ d^L)$$
$$= [\text{prob}_W(d^L_i,\ d^R_+) + \text{prob}_W(d^L_i, d^R_-)] \times$$
$$[\text{prob}_W(d^R_j,\ d^L_+) + \text{prob}_W(d^R_j,\ d^L_-)].$$

(Note that the expression '$\text{prob}_W(d^L_i,\ d^R_+)$' denotes the joint probability of the results d^L_i and d^R_+, and not the conditional probability of d^L_i, given d^R_+.) Jarrett calls this second condition 'the completeness condition'; I shall refer to it as 'the condition of value independence'.[9]

Bell's locality condition, as Bell describes it informally, is the same as Jarrett's locality condition. However, the formal condition which Bell and others actually employ in order to derive an inequivalence is not, Jarrett maintains, the locality condition, but the conjunction of the locality condition and the condition of value independence; Jarrett calls this condition 'the strong locality condition'. In formal terms the condition is: for all states W, and all measured values $d^L_{i=+,\ -}$, $d^R_{j=+,\ -}$, and instances o^L, o^R where no L or R measurements are made,

$$\text{prob}_W(d^L_i,\ d^R_j) = \text{prob}_W(d^L_i;\ o^R) \times \text{prob}_W(d^R_j;\ o^L).$$

This condition requires that the result of a measurement on a L-particle should be statistically independent of both the setting of the R-instrument and the result of measurement on the paired R-particle.[10] Quantum mechanics clearly fails to satisfy the strong locality condition. The failure, moreover, is due either to the failure of the locality condition or to the failure of the condition of value independence. Since the failure of the locality condition may involve a violation of the theory of relativity, it is appropriate, Jarrett suggests, to put the blame for the failure of the strong locality condition on the condition of value independence. The proposal is that quantum mechanics violates the condition of value independence, but not the locality condition.[11] Can this proposal be upheld?

Although they are not logically equivalent, failure of Jarrett's locality condition would strongly suggest a failure of Bell's locality condition. Does quantum mechanics satisfy Jarrett's locality condition? Consider formula (1) (p.182 above) and the case in which failure of the locality condition, if it occurs at all, is most likely, viz. where the angles concerned are $AB = BC = 60°$, $AC = 120°$. In this case it is obvious that the value $A+$ for a measurement on a L-particle (A^L+ for short) is not statistically independent of the value C^L+, since prob(A^L+) does not equal prob(A^L+; C^R+), the former having the value $\frac{1}{2}$ and the latter the value $\frac{3}{4}$. We cannot simply assume, however, that the lack of statistical independence manifested in this case is due to a failure of the locality condition, i.e. to the fact that prob(A^L+) varies depending upon whether or not we arrange to measure the spin-component of the paired R-particle in the direction C. We cannot make this assumption, since prob(A^L+) may vary, being greater if the value of C^R is $+$, and lesser if the value is $-$; and the variation may be such that prob(A^L+; o^R) may equal prob(A^L+; C^R), the probability that A^L has the value $+$, given that the C spin-component of the R-particle is measured. The latter situation does in fact obtain:

$$\text{prob}(A^L+;\ C^R) = \text{prob}(A^L+,\ C^R+) + \text{prob}(A^L+,\ C^R-)$$
$$= \tfrac{1}{2}\sin^2(\tfrac{1}{2}h_{AC}) + \tfrac{1}{2}\cos^2(\tfrac{1}{2}h_{AC}) = \tfrac{3}{8} + \tfrac{1}{8} = \tfrac{1}{2};$$

hence, prob(A^L+o; R) equals prob(A^L; $+C^R$). Thus the statistical frequency of the measured value A^L+ does not vary depending upon whether or not we measure the spin-component of the R-particle in the direction C. The result A^L+ is statistically independent of the setting of the R-instrument. The locality condition is satisfied, moreover, even when the L- and R-instruments are set in the same direction. Say the setting is A; then

$$\text{prob}(A^L+;\ o^R) = \text{prob}(A^L+;\ A^R) = \text{prob}(A^L+;\ A^R+) +$$
$$\text{prob}(A^L+;\ A^R-)$$
$$= o + \tfrac{1}{2}$$
$$= \tfrac{1}{2}.$$

Since the locality condition is satisfied in this case, we have no reason to think that quantum mechanics violates Bell's locality condition.

The quantum-mechanical formula (1) does not entail a violation of the locality condition. Does it involve a violation of the condition of value independence? Obviously it does, in view of the correlations between the results of pairs of measurements in the same direction: if the condition of value independence were satisfied, then prob(A^L+, A^R+) should equal $\frac{1}{4}$, whereas according to quantum mechanics prob(A^L+, A^R+) equals o. The value-independence condition is clearly violated also when measurements are made in different directions. If the condition held in this case, we should have

$$
\begin{aligned}
\text{prob}(A^L+,\ C^R+) &= \text{prob}(A^L+;\ C^R) \times \text{prob}(C^R+;\ A^L)\\
&= [\text{prob}(A^L+,\ C^R+) + \text{prob}(A^L+,\ C^R-)] \times\\
&\quad\ [\text{prob}(C^R+,\ A^L+) + \text{prob}(C^R+,\ A^L-)]\\
&= (\tfrac{3}{8}+\tfrac{1}{8}) \times (\tfrac{3}{8}+\tfrac{1}{8})\\
&= \tfrac{1}{4}.
\end{aligned}
$$

According to quantum mechanics, however, $\text{prob}(A^L+,\ C^R+)$ equals $\tfrac{3}{8}$, and not $\tfrac{1}{4}$. The values A^L+ and C^R+, then, are not statistically independent, owing to the failure of the condition of value independence.

How is the failure of the value-independence condition to be construed? In the case where measurements on the L-wing and the R-wing are made in the same direction, the failure can be regarded as being due to ontic correlations between the pre-existing values of the observables: for any direction d, the value d_i^L depends on the value d_j^R, and vice versa. This explanation, however, cannot possibly be given in the case where measurements on each wing are made in different directions. For certain angles AB, AC, BC, pairs of values $(A+,\ C+)$ should be obtained more frequently than pairs $(A+,\ B+)$ – indeed up to three times more frequently. This state of affairs, however, cannot be explained as being due to ontic correlations between the pre-existing values of the observables, as being due, that is, to pairs of pre-existing values of the types $(+\ +\ -,\ -\ -\ +)$ and $(+\ -\ -,\ -\ +\ +)$ occurring more frequently than pairs of the types $(+\ -\ +,\ -\ +\ -)$ and $(+\ -\ -,\ -\ +\ +)$. We can assign probabilities to pairs of the types $(+\ +\ -,\ -\ -\ +)$ and $(+\ -\ -,\ -\ +\ +)$ in such a way as to satisfy the relation $\text{prob}(A^L+,\ C^R+) > \text{prob}(A^L+,\ B^R+)$, but we cannot at the same time satisfy the relation $\text{prob}(A^L+,\ C^R+) > \text{prob}(B^L+,\ C^R+)$. If, for example, we assign the probability $\tfrac{3}{8}$ to the pair $(+\ +\ -,\ -\ -\ +)$, and the probability o to the pair $(+\ -\ -,\ -\ +\ +)$, then the pair $(+\ -\ +,\ -\ +\ -)$ is assigned the probability $\tfrac{1}{8}$, and so the relation $\text{prob}(A^L+,\ C^R+) > \text{prob}(A^L+,\ B^R+)$ is satisfied; but at the same time the relation $\text{prob}(A^L+,\ C^R+) > \text{prob}(B^L+,\ C^R+)$ is not satisfied, since the results $(B^L+,\ C^R+)$ belong to pairs of the type $(+\ +\ -,\ -\ -\ +)$, to which the probability $\tfrac{3}{8}$ has been assigned. A similar argument holds for every other relevant assignment of probabilities to the various pairs. The quantum-mechanical correlations between pairs of measured values, then, cannot universally be understood as being due to ontic correlations between the pre-existing values of the observables. How, then, are they to be understood? How is the fact that the values A^L+ and C^R+ are not statistically independent to be explained? It is difficult to see how this fact can be explained without invoking the notion of some sort of action at a distance.

Jarrett seems to assume that the notion of some sort of action at a distance needs to be invoked only if we give up the locality condition, and not if we abandon the condition of value independence, for he considers the notion of action at a distance only with respect to the locality condition. Yet, since the failure of the condition of value independence cannot be explained as being due to ontic correlations between the values of the paired observables, the suggestion that the failure is due to some sort of non-local action must be treated seriously. Can the failure be explained without having to invoke the notion of some sort of non-local action?

9.4
Value independence and separability

It might be thought that Howard's interpretation of the significance of Bell's argument provides an answer to the latter question, a solution to the problem of construing the failure of the condition of value independence which does not resort to the notion of some sort of action at a distance. Howard replaces Jarrett's condition of value independence with a condition which he calls 'the separability condition', which is to the effect that two spatially separated objects have their own separate states. He re-formulates Jarrett's strong locality condition as a conjunction of the locality condition and the separability condition. Bell-type inequalities, he maintains, can be derived from the strong locality condition thus construed. Now quantum mechanics does not satisfy the separability condition: it treats a pair of objects which have previously interacted with each other, not as two distinct objects each with its own distinct state, but as a single object with a single, indivisible state. Once two objects have interacted with each other, the state of the system which they jointly compose cannot be described by a state vector which is a product of the state vectors of the component objects; the state vector of the composite system cannot be divided into distinct factors which can be assigned separately to the individual component objects. In Howard's view the reason why quantum mechanics does not satisfy the strong locality condition is that it does not satisfy the separability condition; it does not assign separate states to particles which make up an EPR pair. The satisfaction of the strong locality condition entails that spatially separated objects have their own separate states; and since particles which constitute an EPR pair do not have separate states, they do not satisfy the separability condition, and hence do not satisfy the strong locality condition. The failure of the condition of value independence, then, is due to the failure of the separability condition. This being the case, there is no need to invoke the notion of some sort of action at a distance in order to explain the curious quantum-mechanical statistics. Since an EPR pair constitutes a single object, there can be no question of any interaction between the L-particle and the R-particle.[12] The quantum-mechanical statistics should not be seen as something in need of explanation; they should be seen, rather, as reflecting the quite different conditions of identity of quantum-mechanical objects. Howard's proposal amounts to a radical change in our conception of what constitutes a single object: spatial separation should no longer be regarded as a sufficient condition of the identity of an object.[13]

Does Howard's interpretation really solve the problem of construing the failure of the condition of value independence? His thesis is that two objects which form an EPR pair do not have separate states of their own. It is not clear exactly how this thesis is to be understood. This thesis can be taken in two main ways. First, to the effect that objects which do not satisfy the separability condition are either (*a*) not numerically distinct, but rather numerically identical, or (*b*) not distinguishable parts of the whole which they jointly comprise. On either reading, however, Howard's thesis does not explain the failure of the value-independence condition; for the assumption that observables at the L-wing and the R-wing (and surely these are distinguishable) belong to a single indivisible object with an indivisible state does not entail that the measured values of these observables are not statistically independent, and hence does not in itself explain the failure of the value-independence condition. Thus construed, failure of the separability condition is not in itself sufficient to generate the quantum-mechanical inequality (1); some factor in addition to non-separability is required.

If, however, the requisite additional explanatory factor is the failure of some sort of locality condition, then Howard's thesis may be intended to throw light on this failure, for it may make the notion of action at a distance less puzzling. Howard states that if two objects are not separable, then there can be no interaction between them, since they are not really *two* objects at all, and an interference with one of them is *automatically* an interference with the other.[ib] On reading (*a*), whatever affects one member of an EPR pair *ipso facto* affects the other, since the two are numerically identical; and on reading (*b*), since only *one* object is involved, it is not surprising that what we do at the *L*-wing should have an immediate effect at the *R*-wing. This is not to say that the putative effect involved here is local, but merely to say that there is nothing puzzling about its being non-local. On this first reading, then, Howard's thesis, does not enable us to dispense with the notion of some sort of non-local action; what it does, rather, is to help make this non-local factor seem less puzzling.

On the second way of construing Howard's thesis, objects which do not satisfy the separability condition do not have well-defined states of their own: considered on their own (separately), such objects have no definite states. On this reading, failure of the separability condition entails failure of the intrinsic-properties theory; and hence Howard's thesis is equivalent to the Bohrian response which I am about to discuss.

9.5
The Bohrian response to the Bell–Wigner argument

What would Bohr's response have been to the problem of construing the failure of the condition of value independence? As we saw in Section 8.7, Einstein thought that Bohr rejected the principle of independent existence, according to which the real states of spatially separated objects are independent of each other. I take the latter principle to be logically equivalent to the condition of value independence. Did Bohr reject this principle? I believe he did: he held that what properties, and hence what state, can meaningfully be attributed to an object depends upon the prevailing experimental conditions, and that once an object has interacted with another object in an appropriate way, knowledge of the state of the former can be inferred from our knowledge of the state of the latter. As Bohr clearly saw, one of the two particles of an EPR pair may be treated as an object and the other may be treated as an instrument; after they have interacted, the states of the two particles are correlated in the same way as the states of the object and the instrument are correlated by the measurement interaction; hence a suitable observation of the one particle (the instrument) enables us to infer the correlated value of an observable of the other (the object). Moreover, just as the process of measurement does not create the value that is measured, so the observation made on one particle (the instrument) does not create the correlated value of the other particle (the object). Hence EPR-type correlations do not involve any action at a distance. Bohr, then, wished to reject the condition of value independence and to uphold what might be called 'the principle of local action', which forbids action at a distance. But can Bohr legitimately do both of these things while maintaining the objective-values theory?

Indeed, he can: Bohr can explain the failure of the condition of value independence as being due to ontic correlations between the measured values of paired observables, and he can do this even in the recalcitrant case where the measured values are A^L+, C^R+, B^R+. The reason for this is that if we do not presuppose the intrinsic-properties theory, as Bohr does not, it is possible to maintain both the objective-values theory and the principle of local action. If we assume only the objective-values theory, we are not entitled to assume that observables which are not measured have pre-existing values; it may be that such observables cannot meaningfully be said to have values at all. If we do not assume the intrinsic-properties theory, we cannot make use of the inequality (2), and it is easy to see that the Bell–Wigner inequality (1) does not necessitate a violation of the principle of local action. If we assume that observables which are not measured cannot meaningfully be said to have pre-existing values, then the probability expressions can be represented as follows:

$$\text{prob}(A^L+, \ C^R+) = \text{prob}(+*-, \ -*+)$$
$$\text{prob}(A^L+, \ B^R+) = \text{prob}(+-*, \ -+*)$$
$$\text{prob}(B^L+, \ C^R+) = \text{prob}(*+-, \ *-+)$$

The expression 'prob$(+*-, \ -*+)$' signifies the probability that an L-particle has the measured values $A+$, $C-$, and the paired R-particle the measured values $A-$, $C+$, the asterisks marking the places where no values can meaningfully be assigned. In this expression it is presupposed that measurement of the spin-component of one particle in a certain direction constitutes also a measurement of the spin-component of the paired particle in that same direction. If this assumption is not made, then the expression 'prob$(A^L+, \ C^R+)$' is represented as 'prob$(+**, \ **+)$', and similarly in the other cases. I shall present the argument only in terms of the former style, since the argument goes through just as well in terms of the latter style. The quantum-mechanical inequality (1) can now be expressed as follows:

$$\text{prob}(+*-, \ -*+) > \text{prob}(+-*, \ -+*) + \text{prob}(*+-, \ *-+).$$

Now the fact that the statistical frequency of pairs of the type $(+*-, -*+)$ may be very much greater than that of pairs of the type $(+ -*, - +*)$ does not require an explanation in terms of some sort of action at a distance: the observed statistical frequency may simply reflect the actual distributions of the values which the measured observables possess immediately before they are measured; pairs of the type $(+*-, -*+)$ may simply occur more frequently than pairs of the type $(+ -*, - +*)$ and $(*+ -, *- +)$. Thus it is possible to reject the intrinsic-properties theory and to maintain both the objective-values theory and a locality condition. This is exactly what Bohr does. Bohr, then, is able to provide a solution to the problem of construing the failure of the condition of value independence, a solution which does not require him to abandon the principle of local action. On this solution it is possible to explain the failure of the condition of value independence as being due to ontic correlations between the values of measured observables. What makes this explanation possible is the rejection of the intrinsic-properties theory right at the outset; if that theory is presumed, then, as we saw at the end of Section 9.3, no such explanation is possible. Apart from Fine's 'prism' theory, the Bohrian solution to the problem is the only one I know of which enables us to maintain both the objective-values theory and the principle of local action.

Let us recall the question which was shelved at the end of Chapter 6 (p.133 above): Does Bell's argument pose any serious problem for the objective-values theory? As we have seen, the Bell–Wigner argument does not impugn the intrinsic-properties theory as such, but rather the conjunction of the intrinsic-properties theory, the objective-values theory, Bell's locality condition, and the condition to the effect that the measurement statistics exactly match the distribution of the pre-existing values (the matching condition). We may therefore retain both the objective-values theory and Bell's locality condition, provided we abandon the intrinsic-properties theory (as Bohr does). Alternatively, we may retain both the intrinsic-properties theory and the objective-values theory, provided we dispense with Bell's locality condition or the matching condition. We are required to relinquish the objective-values theory only if we retain both the intrinsic-properties theory and Bell's locality condition. The Bell–Wigner argument, then, poses no insuperable problem for the objective-values theory.

These considerations show that Einstein's position cannot be sustained: Einstein wished to uphold the intrinsic-properties theory, the objective-values theory, and a locality condition. Which of these conditions would he have chosen to relinquish? No doubt the locality condition, since he held the latter to be more dispensable than the intrinsic-properties/objective values conditions. For Bohr, by contrast, the Bell–Wigner argument poses no problem, since he does not hold the intrinsic-properties theory.

It is the mark of a good theory that it can explain some newly discovered fact entirely from its own resources, without the need for modification. Bohr's interpretation of quantum mechanics has exactly this virtue: without modification of any of its basic assumptions, it can explain the novel and refractory data which Bell has brought to light. Einstein's interpretation, by contrast, can explain these data only if some of its basic assumptions are dispensed with. The sequel to the great debate, then, favours Bohr rather than Einstein.

It might be thought that Bohr's notion of wholeness throws light on the nature of the quantum-mechanical statistics concerning EPR pairs. It is true that two objects which form an EPR pair constitute a Bohrian whole, since, theoretically at least, one of them may be treated as an object and the other may be treated as an instrument. The negative correlations between the spin-components of two such objects are instances of the sort of correlations which characterise the measurement process. To this extent the notion of wholeness has some explanatory value in this context. Again, an EPR pair together with the macroscopic apparatus which is used to make a measurement on it forms a Bohrian whole. I cannot see, however, that the notion of wholeness helps to explain the nature of the statistical correlations between results of measurements on paired observables in different directions. Perhaps if it were thought necessary to invoke the notion of an action at a distance to explain these correlations, the concept of wholeness might be fruitful, since the idea of an organic interconnection between the paired objects might help to cast light on the matter. But what light exactly?

How good is the Bohrian explanation of Bell's curious data? Ought we to accept it? It has several advantages: it preserves the objective-values theory and the principle of local action. Its main disadvantage is that it sacrifices the intrinsic-properties theory. That theory may of course be simply a piece of outmoded metaphysics that we are better off without. What are the alternative explanations? If we wish to preserve the intrinsic-properties theory, we must either adopt Fine's 'prism' theory or relinquish Bell's locality condition. The latter alternative has received strong support in some quarters. Jean-Pierre Vigier, for example, postulates the real physical existence of de Broglie waves, which mediate an action at a distance caused by a phase-like disturbance of the quantum potential. Vigier's theory is similar in many respects to Bohm's hidden-variables theory.[14] Such theories have problems of their own: to mention only one, it is a moot question whether abandonment of Bell's locality condition can be reconciled with the theory of relativity.[15] Compared with its rivals, Bohr's interpretation is no more problematical; rejection of the intrinsic-properties theory plays greater havoc with our metaphysical intuitions than does rejection of the principle of local action; by the same token, however, it is all the more adventurous. I shall explore the philosophical foundation of Bohr's interpretation in the following chapter. But first, by way of contrast, let us look briefly at Einstein's general philosophy of physics.

9.6
Einstein's philosophy of physics

Einstein gave a fairly full account of his philosophy of science in his paper of 1936, 'Physics and Reality'.[16] He continued to hold the views expressed there, with little modification, until the end of his life.

The aim of physical science is, Einstein holds, to help us make sense of our perceptual experience, to co-ordinate our sense impressions by means of general concepts.[17] General concepts, he maintains, are not abstractions from sense data; they are 'free creations' of the mind.[18] About the process of creation, little can be said.[19] We form general concepts by selecting certain constantly recurring complexes of sense impressions and attributing to them a 'meaning', which is to a large extent independent of the impressions from which they are formed. The concept, however, is not identical with any particular set of sense impressions. The meaning of the concept is the real object which it signifies. It is the independence of the concept from the specific sense impressions which constitutes the 'real existence' of the object.[20] Our conception of reality is largely a product of the mind:

> "Being" is always something which is mentally constructed by us, that is, something which we freely posit (in the logical sense). The justification of such constructs does not lie in their derivation from what is given by the senses . . . [but] in their quality of making intelligible what is sensorily given . . .[21]

Our conception of objective reality is not derived immediately from the senses; it is a product of the mind's organisation of sensory data by means of concepts which the mind itself produces. The similarity here between

Einstein's theory of concepts and Kant's doctrine is unmistakable; indeed, it was noted by Einstein himself. He differs from Kant, however, in holding that the most basic concepts, the categories, are not *a priori*, conditioned by the nature of our understanding; they are, like any other concepts, free conventions, however pervasive and established in our thinking.[22] On this issue the Kantian influence on Einstein's thinking was tempered by the influence of Henri Poincaré and the neo-Kantians.

Einstein conceives of a conceptual scheme as a sort of ordered structure, at the lowest level of which are the 'primary' concepts, i.e. empirical concepts directly associated with complexes of sense-data. Resting upon these are higher-order concepts which derive their meaning partly from connections with the primary concepts. The highest level of the conceptual schemes of physical theories comprise concepts which have remote and indirect connections with the primary concepts.[23]

Whereas common-sense conceptual schemes contain a diversity of concepts, scientific conceptual schemes seek the greatest possible economy and simplicity, i.e. 'logical unity', in their conceptual foundations. The aim of science is 'to discover the greatest possible number of empirical facts by logical deduction from the smallest possible number of hypotheses or axioms'.[24] As the conceptual basis of physics becomes simpler and more homogeneous, it also becomes further removed from experience and hence more difficult to create and more prone to error.[25] Although Einstein was inclined to doubt whether the aim of achieving the greatest possible conceptual unity is attainable, he believed that the aim should be maintained as a regulative principle.[26] He was greatly troubled by the lack of conceptual unity in the foundations of contemporary physics, at the heart of which lie the two very disparate concepts of discrete mass and continuous field, according to which energy is conceived of as being now discontinuously, now continuously, distributed in space. The difficulty in uniting quantum theory with relativity theory was, he believed, due largely to a lack of conceptual unity:[27] what was needed was a unified conceptual scheme subsuming the schemes of mechanics and electrodynamics, and uniting the notions of the electromagnetic and gravitational fields.[28] Here Einstein differs from Bohr, for whom conceptual homogeneity *per se* was not one of the main goals of science.

Just as Einstein rejects the abstractivist doctrine of concept formation, so also does he reject the inductivist view of theory formation: 'A theory can be tested by experience, but there is no way from experience to the setting up of a theory.'[29] Like Popper he holds that theory formation is an intuitive, constructive process: scientific theories are, like concepts, 'free creations' of the mind. The fact that two quite different theories can explain the same facts indicates that the relation between fact and theory is not deductive.[30] He is particularly critical of the view of Mach, Kirchhoff and Ostwald that physics ought only to make use of concepts that can be inductively derived from experience. He rejects Mach's preference for 'phenomenological' theories. The ultimate aim of science, viz. the greatest possible conceptual unity, cannot be attained through phenomenological theories.[31] This is not to say that he denies the effectiveness of some phenomenological theories, such as the phenomenological theory of heat.

Einstein distinguishes between two sorts of physical theories: principle theories and constructive theories. Principle theories on the one hand are formed by an analytic method; they are 'suggested' by empirical data, though not inductively derived from them. Such theories formulate general principles which empirical phenomena are required to satisfy. Phenomenological thermodynamics and the special theory of relativity are principle theories: thermodynamics rules out the possibility of a *perpetuum mobile* of the first and second kinds, while relativity theory requires the laws to be covariant with respect to the Lorentz transformations. The advantages of principle theories are 'logical perfection and security of the foundations'.[32] Constructive theories on the other hand are formed by a synthetic process: they are based upon a relatively simple conceptual and formal scheme which is an imaginative construction. The molecular theory of gases, for example, is a constructive theory: it explains the properties of gases, such as viscosity, diffusibility, heat conductivity etc. in terms of the mechanical laws governing the motions of molecules.[33]

There is a fairly close resemblance between Einstein's philosophy of science and Popper's, and perhaps an even closer similarity between Einstein's and William Whewell's, which, curiously enough, seems to have gone unnoticed. Since most of the similarities between Einstein's philosophy and Whewell's philosophy apply also to Popper's, and since the latter is the better known, I shall comment only on the similarity between Einstein's ideas and Whewell's.

Einstein and Whewell both stress that scientific knowledge contains an empirical element ('sense-impressions' for Einstein, 'facts' for Whewell), and a conceptual element ('concepts', 'ideas').[34] Where Einstein says, 'We represent the sense-impressions as conditioned by an "objective" and by a "subjective" factor',[35] Whewell says, 'The combination of the two elements, the subjective or ideal, and the objective or observed, is necessary, in order to give us any insight into the laws of nature.'[36] Both Einstein and Whewell, moreover, hold that there is no firm distinction between sense-impressions and concepts, between facts and theories.[37] Heisenberg recalls that Einstein's remark that it is the theory which determines what can be observed was an important influence on his thinking at the time when he put forward the uncertainty principle.[38]

Einstein and Whewell hold very similar views about the formation and testing of theories. Both reject the inductivist account. For Whewell theory construction begins with the suggestion of an idea, a unifying conception which 'colligates' a set of facts. Kepler's idea of elliptical planetary orbits for example was not an inductive generalisation of empirical data; it was, rather, 'suggested' by the data and superimposed upon them.[39] Whewell and Einstein hold the hypothetico-deductive view of scientific theories: theory is tested by comparing with experience the conclusions derived from it; what constitutes experience of course is itself partly determined by theory.[40]

Both Whewell and Einstein are agreed that the better a scientific theory, the greater its conceptual unity and formal simplicity, and the wider and more diverse the range of empirical phenomena explained by it (Whewell's 'consilience of inductions').[41] Einstein considered it as one of the greatest merits of his theory of relativity that it is a simple generalisation of hypotheses previously thought independent of one another.[42]

Einstein and Whewell differ, however, over the alethic status of theories. Whereas Whewell allows that certain theories may attain the status of necessary truth, Einstein rejects the view that any theory is necessarily true. Moreover, whereas Einstein holds that no concepts are *a priori*, Whewell maintains that the concepts of space, time and causation are *a priori*. In this respect Einstein departed further from the Kantian path than Whewell: his philosophy of science nevertheless is firmly within the neo-Kantian tradition that was so deeply influential in continental philosophical circles in the early years of this century.

It is obvious from this account of Einstein's general philosophy that it is quite wrong to describe Einstein as some sort of naive realist: his realism is of a sophisticated sort, containing as it does elements of neo-Kantian conceptual idealism. Einstein holds that the subject-matter of a physical theory is physical reality as it is independently of our observation of it, but our conception of physical reality is determined not by observation alone but also by our conceptual scheme, which is the product not of the senses but of the mind. Observation is theory-laden, mediated by our concepts: whereas the content of experience derives from the senses, its form derives from the mind. Physical reality cannot be 'read off' directly from our experience, since our experience is steeped in the concepts which the mind spontaneously creates. The notion of reality is primarily an explanatory concept, an idea which we create in order to make sense of our experience. No *particular* concept – not even the concept of physical reality – is absolutely binding. What is real, then, cannot be determined *a priori*. For example, whether or not an individual radio-active atom disintegrates at a definite instant is a question that is meaningful only in the light of a particular conceptual scheme.[43] Einstein does not assume that we can say *a priori* or independently of some chosen conceptual scheme what the characteristics of physical reality must be. He does not assume that an electron must be *a priori* to have a definite simultaneous position and momentum; he believes, however, that given the conceptual scheme of quantum mechanics the question of the physical reality of the attribute of definite simultaneous position and momentum can meaningfully be asked. Unlike Bohr he can see no compelling reason for regarding the classical conceptual scheme as having a limited applicability at this point. Einstein was committed to certain concepts – such as the notions of realism (all objects have definite, observer-independent properties at all times), determinism (all events fall under strictly deterministic laws) and locality (all events have local causes) not because he regarded these as *a priori* in any sense, but because he held that notions, like these, which again and again have proved useful in making sense of our experience should be jettisoned only if there is a compensatory gain, not just in the empirical scope of our theories, but also in theoretical unification – a condition which, he thought, had not been met where the concepts of realism, determinism, and locality were concerned. As we shall see, Einstein's general philosophy of science is much closer to Bohr's than either of these two great physicists realised: Einstein's is less realist than Bohr suspected, and Bohr's is less idealist than it seemed to Einstein.

Bohr's philosophy of physics

Now that Bohr's philosophy of quantum physics has been surveyed in detail, it is time to provide a general description of his philosophy that illuminates the details. The question which I wish to address, then is: what is the general philosophy, or philosophical point of view, that underlies Bohr's philosophy of physics? Can his philosophy be neatly categorised in terms of any well-known philosophical position? Apropos his philosophy of physics in particular, the question of primary interest is: to what extent, if any, is Bohr's philosophy of physics realist? No simple answer to this question can be given: indeed, certain realist and non-realist components are subtly interwoven in his philosophy, and to unravel them we shall need a fairly refined conception of what constitutes a realist interpretation of physics.

10.1
Realism in the interpretation of physics

The basis of realism is the view that statements of a certain class predicate properties of real objects, the truth-values of these statements being determined by reality as it is independently of how it appears to us to be. This notion is captured in the following characterisation of what constitutes the adoption of a realist interpretation of a physical theory.

(1) The physical theory is intended to explain, and not merely to predict, phenomena in terms of the postulated physical reality which underlies them. Hence certain sorts of sentences belonging to the theory may be genuinely propositional, i.e. have truth-values, these truth-values being determined by physical reality independently of our knowledge of it.

(2) Certain theoretical terms purport either to denote real physical entities which may not be directly observable by means of the senses, or to express real physical properties.

These two propositions comprise the fundamental tenets of realism: they convey the two main points, viz. that a theory has an essential reference to physical reality, the existence of which is independent of human knowledge and perception, and the theory is true or false of that reality independently of what we know or perceive. Many realists adopt a further proposition:

(3) The main aim of physics is the construction of explanatory theories which approximate as closely as possible to the truth about physical reality.

Some go further, maintaining:

(4) There are grounds for thinking that as physics develops it provides theories which are in closer approximation to the truth about physical reality than previous theories.[1]

According to proposition (4), acceptance of a theory does not require that one should believe the theory to be true or even approximately true; that requirement belongs to a much stronger form of realism again.

As a general theory of the interpretation of physics, realism has come under considerable attack of late. Before I examine Bohr's position, it will be useful to make a brief survey of some of the main lines of criticism. I shall not attempt to do justice to the complexity of the issues involved, or to indicate all the defences that are open to the realist.

First, *the arguments against a realist conception of truth*: proposition (1) seems to call for a realist conception of truth, one in which the notion of reality plays a substantive role; but a realist theory of truth faces difficulties. The correspondence theory, for example – to the effect that the truth of a statement consists in a relation of correspondence between the statement and a fact – faces the difficulty of explaining what a fact is, and what the relation of correspondence consists in – difficulties which are too well known to need rehearsing here. The root of the problem is the faulty conception of truth as being analogous to the relation between a picture and the object which is pictured. A more modern theory states that the truth of a statement consists in a relation of reference between certain of its linguistic expressions and items of reality. On this view a statement is true if and only if its subject term denotes or refers to some real item which belongs to the extension of the predicate term. The notion of reference, however, is just as puzzling as the notion of correspondence. Some philosophers despair of making the notion intelligible at all. W.V.O. Quine, for example, has argued that the referent of a term is not determinately fixed, and consequently reference is inscrutable in the sense that we can have no sure knowledge of what a speaker of a foreign language, or even of our own language, is referring to. Quine's complex and highly controversial argument calls in question the very notion of a determinate relation of reference between language and the world.[2]

The idea of a determinate relation of reference between language and reality has been attacked from a different angle by Hilary Putnam. Generalising a consequence of the Löwenheim–Skolem theorem to the effect that an axiomatic formulation of set theory may have many different sorts of models, denumerable and non-denumerable, Putnam argues that an ideal theory of the world – one which is compatible with all possible observations and satisfies all theoretical and operational requirements – has different models. Since these models satisfy all epistemic conditions, there is no non-arbitrary way of selecting one of them as the unique, intended model, i.e. as representing the world; and hence it is impossible to fix a unique, determinate reference for the terms of such an ideal theory. The problem, then, is: since there are alternative possible models of the theory, how do we know what the right model, viz. reality, is?

Putnam maintains that reference cannot be fixed by our mind's selecting reality as the intended model, since that would require the unintelligible notion of our mind's having access to reality that is unmediated by the theory whose reference is in question. Putnam also rejects the view that reference is determined by a causal relation between an object and the language user, the intended referent being that object which has played the primary causal role in the user's acquisition of the correct use of the referring expression. The causal account of how reference is fixed will not do here, since it presupposes realism in assuming that there is a single reality which is one of the terms in the causal relation, when it is just this assumption that the argument from the multiplicity of possible models calls in question.[3]

Some argue that it makes no sense to say that the reference of our terms, or the truth-value of our statements, is determined by a mind-independent reality, i.e. a reality the existence of which is independent of human thought or knowledge of it, since this conception of reality is vacuous. The argument, a venerable one, is that on the realist conception we could have knowledge of the referent of a term or the truth-value of a statement only if we could compare the term or the statement with mind-independent reality; but this cannot be done, since we can grasp reality only through the medium of concepts, and so we can compare our thoughts only with conceptualised reality. The only reality to which we have access is reality-as-it-appears-to-us, which *eo ipso* is not mind-independent. No significant meaning, therefore, can be given to the notion of mind-independent reality.[4] This argument is flawed by the fallacy of equivocation. In one sense, to say that some item is mind-independent is to say that the item exists independently of the existence of any human mind; in another sense, to say that some item is mind-dependent is to say that it cannot be apprehended by the human mind in a way that is unmediated by some conceptual scheme. The fact, however, that some item is not mind-independent in the latter sense does not entail that it is not mind-independent in the former sense; and it is the former sense that is at issue in the realist conception of truth. It is true, but trivial, that we cannot conceive of an unconceptualised reality, since we cannot conceive at all without employing concepts; but, for all the above argument shows, the reality that we do conceive of may exist independently of any human conception of it. Moreover, the idea that, given a realist conception of truth, we acquire knowledge of the truth-value of a statement only by comparing the statement with unconceptualised reality is a grotesque confusion. Given the truth-value which reality determines, we get to know this truth-value, if knowledge is possible at all, by coming to believe that the statement is true when in fact it is, and by having reasons for our belief which provide adequate justification for it. If one thinks of concepts as screens, whether transparent or opaque, or as pictures of reality, then it is reasonable to think that we cannot tell whether our concepts faithfully present reality to us, since, in this case, we cannot compare the picture with the reality that is pictured. But if one thinks in this way, why should one think that our concepts present an obscure or distorted image of reality to us? The source of the confusion is the conception of concepts as metaphorical pictures.

Those who accept some version of the above argument are inclined to hold that truth is not a relation between language and extra-linguistic reality, but rather a relation between a statement and a certain designated set of statements. Some relativists, such as Kuhn and Feyerabend, maintain that there is no unique, privileged set of statements, so that truth, if the concept has any meaning at all, is relative to the conceptual schemes of actual, particular communities of thinkers at a particular time. On this view, truth is objective in the sense that it does not depend upon what individual human beings think. Others hold that truth is relative to a unique, privileged set of statements, and is objective in the sense that it does not depend upon what is generally accepted by any actual community of thinkers. C.S. Peirce, for example, held that truth is a relation of coherence with the set of statements upon which the belief of a community of ideally rational inquirers would converge if their inquiry were pursued indefinitely. Putnam holds a similar view: truth is what we would be justified in believing under epistemically ideal conditions. On this view truth does not depend upon what any individual or group of individuals thinks here and now, but it does depend upon what would be thought by ideally rational cognitive (and presumably human) beings.[5] Putnam calls his theory 'internal realism', no doubt because what determines truth is not external to thought; but it would be better to call it 'conceptual idealism' as Nicholas Rescher calls his view, since truth is taken to depend not upon reality but upon thought (ideas).

Second, *the argument from the underdetermination of theory by data*: it can be argued that, given any physical theory T, it is possible to construct a rival theory T' which is such that (*a*) T and T' are logically incompatible in the sense that each contains a theoretical statement the negation of which is contained in the other; and (*b*) T and T' are evidentially equivalent in the sense that they are empirically equivalent (i.e. have the same empirical consequences) and fare equally well with respect to all other criteria of theory assessment. A theory is said to be underdetermined in this sense if it is possible to construct a rival theory that satisfies (*a*) and (*b*). Since, the assertion is, every theory is underdetermined in this sense, we can have no grounds for believing that any particular theory is true or closer to the truth than any other theory. It might be thought that only one of two such rival theories could be coherently integrated into a total theory of reality, a theory that is compatible with all possible observations; in which case, if we possessed such a total theory, we could make a rational choice between T and T'. But again, the argument goes, it is possible to construct a logically incompatible and evidentially equivalent total theory of reality; and so the problem of underdetermination remains. Now with respect to the statement S, over whose truth-value T and T' disagree, there are two possible responses. One could maintain that S is determinately true or false, even although it is impossible to tell which it is – the 'ignorance response' as W. Newton-Smith calls it. On this response, there may be matters of fact which are totally inaccessible to us. In this case physical theory cannot capture, or at least be known to capture, all the truth about physical reality. This response requires the abandonment of propositions (3) and (4) in their present form, the reason being that, for any given theoretical statement S, it is possible to construct an evidentially equivalent theory which contains the negation of S; in which case we cannot know what the truth-value of S is, and consequently the aim of physics cannot meaningfully be said to be the approximation to the truth, at least not without major qualification. Alternatively, we could maintain that with respect to S there is no fact of the matter the 'arrogance response'. On this response, since there is no way of rationally assigning S a truth-value, it is pointless to maintain that S has a determinate truth-value, pointless since no empirical fact is explained by this assumption. The arrogance response commits one to abandoning propositions (1) – (4). It is important to note that this version of the argument requires only that it is possible, for any given theory, to construct a rival theory satisfying conditions (*a*) and (*b*); it is not required that we actually possess such a theory. Putnam, like Newton-Smith, favours the arrogance response: the possibility of rival, evidentially equivalent theories is fatal to the view that theories describe reality *per se*; what they describe is a theory-dependent reality, a reality that is internal to thought.[6] The main weakness of this objection to realism is that those who propose it have not done enough to establish that the possibility of such rival theories is a genuine one.[7]

Third, *the argument from the 'pessimistic induction'*, as Newton-Smith calls it: in the light of contemporary physics most empirically successful theories in the past are now taken to be false, and some of their theoretical terms are now thought not to denote real entities or to express real physical properties (think of the notion of the electromagnetic ether). We have no good grounds, then, for thinking that our present physical theories will fare any better in this respect, and so we have no grounds for thinking that our present theories are a closer approximation to the truth than those of 1850, say: the notions of the atom, the electron, the proton, for example, may go the way of the notions of phlogiston and the ether.[8]

This argument undermines propositions (2), (3), and (4). It also threatens an argument which some realists put forward in support of their position, the argument from the empirical progress of physics. As physics develops, it adequately explains and predicts a wider and wider range of phenomena. The only plausible explanation of this fact, it is argued, is that as physics develops it approximates more and more closely to the truth about physical reality. Thus phenomena are as they would be if there were atoms, electrons etc., simply because in reality there are atoms, electrons etc., though our present conception of these may be only partially correct. If atoms, electrons, etc., did not exist, then the empirical success of contemporary physics would be 'miraculous'.[9] The argument from the empirical success of contemporary physics is an argument to the best explanation.

Fourth, *the argument from the incommensurability of radically different theories:* Feyerabend and Kuhn argue that we can have no grounds for thinking that as physics develops it approximates more closely to the truth, since there is no way of comparing radically different theories (Kuhn speaks of 'paradigms'): such theories are incommensurable, and hence we cannot say that one is closer to the truth than another. The argument for this thesis is that (i) the very character of one's sensory experience is affected by the physical theory that one holds, and so those who hold different theories do not perceive the world in the same way. Consequently there is no common, theory-neutral, empirical basis against which different theories can be assessed with respect to their truth content (the radical theory-dependence argument). (ii) The meanings of the terms of a theory are determined by the theory itself: the meanings of the theoretical concepts are functions of the theoretical laws into which they enter. Terms belonging to radically different theories, therefore, must differ radically in meaning; and since reference is determined by meaning, the references of these terms must differ also. Different theories, consequently, cannot be taken to describe the same physical reality or the same aspects of reality, and hence are incomparable (the radical meaning-variance argument). (iii) The very problems posed and the methods for solving them and for assessing proposed solutions are specific to particular theories; and hence, those who hold radically different theories do not agree about what problems should be solved, or methods for solving them, or criteria for assessing solutions; and there is no rational way of resolving this disagreement (the criteria-variance argument).

Defenders of realism reply that (i) while observation is theory-laden it is either not laden with the theory to be tested, or not laden to such an extent that the theory cannot be tested by theory-neutral data: (ii) while the sense of a theoretical term may be partially determined by the theory that contains it, in general there is an overlap between the meanings of a term that is common to different theories, which is sufficient to provide grounds for thinking that the referent of the term is the same; (iii) while the criteria for the selection of problems, solutions, and methods may to some extent differ from theory to theory, this change is genuinely progressive in that as physics develops it provides more and more effective criteria for theory selection.[10]

Propositions (3) and (4) face further difficulties, one of the most pressing of which is that as yet no precise concept of approximation to the truth has been formulated that is logically coherent. Popper's theory of verisimilitude has been shown to be inconsistent, and other attempts to construct a formal theory have met with enormous difficulty. Nevertheless the fact that no adequate formal theory of approximation to the truth is available at present does not entail that the notion is incoherent; it may merely reflect the enormous complexity of the problem.[11]

There are two main alternatives to a realist construal of physical theories: weak and strong instrumentalism, as I shall call them. The weak instrumentalist does not deny that certain sorts of theoretical sentences have truth-values, but he denies that the notion of truth plays any substantive role in our understanding of physics: that notion does not serve to define the aims of physics or explain its progress. Versions of weak instrumentalism have been proposed recently by Larry Laudan and van Fraassen. Laudan maintains that the aim of science is to solve empirical problems, and that theories are devices for producing solutions.[12] Van Fraassen maintains that the aim of science is not truth but empirical adequacy; the aim is to provide a true description, not of physical reality, but of what is observable, to 'save the appearances'; and acceptance of a theory does not involve the belief that it is true, but merely that it is empirically adequate, that it squares with the empirical data. Van Fraassen calls his theory 'constructive empiricism'.[13]

The strong instrumentalist denies that physical theories have truth-values: they have no genuine descriptive or explanatory content at all; they are merely linguistic devices for predicting the course of our sensory experience and for organising it conceptually in an economical way.

My intention in recounting the above arguments is not to endorse the case against realism, but to indicate the nature and extent of the difficulties that face a realist construal of physical theory in general. A realist construal of quantum physics of course faces further difficulties still.

10.2
Bohr and scientific realism

Is Bohr's philosophy of physics realist? In answering this question let us begin with a particularly strong form of realism, scientific realism, which comprises, in addition to propositions (1)–(4) of the previous section, the thesis that current physical theories, providing as they do the best available account of physical reality, should be believed to be true, and since physics contradicts many of our common-sense beliefs, these are to be rejected as false. On this view, atoms and elementary particles exist, but in reality there are no such things as the physical objects – such as tables and chairs – conceived of by common sense.[14]

Bohr, clearly, rejects scientific realism. The common-sense conception of the world of material objects, he holds, is not radically mistaken, the mere appearance of some intelligible, but imperceptible, reality; rather, the common-sense conception is the foundation upon which all objective communication rests. For Bohr the two conceptions of physical reality, the common-sense and the quantum-physical, are not incompatible but complementary, and hence appropriate to different points of view or modes of apprehension. This is the view that is adopted by Strawson, who argues that our conception of the real properties of physical objects is relative to a conceptual point of view, and the common-sense and scientific conceptions are relative to different points of view. Once this relativity to a point of view is recognised, the two conceptions can be seen to be perfectly compatible. Moreover, we could not occupy the scientific point of view if we did not first occupy the common-sense one: 'Science is not only the offspring of common sense; it remains its dependant.'[15] Here Strawson is echoing Bohr, no doubt unwittingly (he does not use the term 'complementary').

A more sophisticated version of scientific realism has been advocated by Michael Dummett. Like Strawson, Dummett denies that the common-sense conception of the world is radically mistaken or incompatible with a scientific conception (though details of the two conceptions may be incompatible). The difference between these two conceptions is that whereas the former is relative to our peculiarly human modes of perception, the latter purports to be independent of these, i.e. to describe things in absolute terms. Moreover, the scientific conception is a refinement of the common-sense conception. We have therefore, no option but to take the scientific conception as revealing what the world is like in itself. The common-sense conception is simply a low-grade theory which has become deeply entrenched in our thinking; but we owe it no special allegiance.[16]

While some of Dummett's views are exactly in line with those of Bohr, his brand of scientific realism is still too robust for Bohr. In particular, Bohr would reject the notion of a description in absolute terms, i.e. in terms that are independent, not only of the perceptual experience of individual observers, but also of human modes of perception in general and common-sense modes of conceiving. Theoretical concepts which are devoid of any perceptual content are for Bohr 'abstractions' or 'idealisations', which we are not justified in interpreting in realist terms. It is for this reason that he stresses the common-sense roots of physics: if theoretical concepts are to have any genuine explanatory force, they cannot lose all connection with ordinary human sense perception, with, that is, our ordinary, everyday concepts. Physical theory, then, cannot be as absolute as Dummett envisages.

Dummett seems even to agree with Bohr on this point, for he states that once science goes beyond some critical level of abstraction, where its concepts lose all connection with ordinary perceptual experience, it can no longer be regarded as providing a description in absolute terms, and a thoroughgoing scientific realism ceases to be tenable. He thinks, moreover, that this critical level has already been passed by quantum electrodynamics, with its concepts of 'colour' and 'strangeness' that are devoid of perceptual content.[17] Bohr would insist that Dummett is inclined to put the critical level of abstraction far too high; for he holds that there is already considerable (though often unrecognised) abstraction even with respect to fairly elementary concepts of physics. If Bohr is right on this point, then it is questionable whether or in what sense there is, as Dummett contends, 'a legitimate notion of a description of an object as it is in itself'.(ib.) If one of the most successful physical theories ever devised cannot be said to provide descriptions of the objects as they are in themselves, then it is doubtful whether scientific realism, even of Dummett's more sophisticated variety, is a tenable general theory of the cognitive status of physics.

It is important to note that realism, as characterised by proposition (1) of Section 10.1, does not require that true statements be regarded as conveying conceptual pictures which faithfully depict the inherent properties of the object as it is in itself, that is, that they are true in an absolute sense. Bohr is sceptical of the view that the human mind, divorced from its perceptual capacities, is able to grasp in thought physical reality as it is in itself. Bohr's attitude towards the conceptual schemes of theoretical physics is analogous to the common-sense realist's attitude to ordinary sense perception. Unlike the naive realist, who holds that the external world is in itself exactly as we ordinarily perceive it to be, the common-sense realist holds that, while we do in fact perceive external objects (as distinct from sense data), these objects in reality are not in all respects exactly as we perceive them to be. Dummett's attitude towards physical theory, by contrast, is analogous to the naive realist's attitude towards ordinary sense perception, in that he holds that physics may accurately describe the world as it is in itself; whereas he rejects naive realism with respect to perception, he entertains it with respect to conception. Dummett's realism, for Bohr, is still too sanguine.

10.3
Bohr and empirical realism

Bohr, then, is not a scientific realist; indeed, he is an empirical realist, holding that the common-sense conception of the physical world is not radically mistaken or contradicted by modern physics. He holds that our sensory experience is of the real world of external objects whose existence is independent of our perception of them. It is this real world, common to all subjects of experience, that is the basis of intersubjective communication. It is empirical realism, and not radical empiricism, that underlies his constant stress on the need to describe the experimental apparatus in ordinary, everyday-language terms. The existence of macroscopic material objects is for Bohr a necessary condition of unambiguous, intersubjective communication. His thinking here has a close affinity with Strawson's theory of the role that the notion of a material object plays in the conceptual scheme in terms of which we think and talk about particulars.

Strawson argues that material objects – bodies – are ontologically basic, i.e. basic within the ontology of our common-sense conceptual scheme. His argument may be briefly formulated as follows. Intersubjective communication in a common language is possible only if speaker and hearer can identify and re-identify the particulars that are being spoken about. Identification can be either demonstrative – which involves directly locating or picking out, by means of sight, hearing or touch, the particular referred to – or descriptive, which involves knowledge of some individuating description uniquely applying to the particular concerned.[18] Now it is a necessary condition of the possibility of descriptive identification and re-identification of particulars that there should be a single system of spatio-temporal relations within which each particular has a unique position, and is uniquely related to every other particular and to the speaker and the hearer, a system comprising one temporal, and three spatial dimensions.[19] This system is constituted by material objects: without these we should have no access to such a system, since the system must be perceivable; and material objects are the only objects that are perceivable, three dimensional, and enduring through time. The spatio-temporal reference system constituted by material objects provides a public reference point, common to speaker and hearer, to which every particular referred to is uniquely related. Since they constitute just such a reference system, material objects are basic from the point of view of particular identification. They are particulars of a sort that can be identified and re-identified without reference to particulars of a different sort; in this respect they are ontologically basic within our conceptual scheme.[20] Basic particulars must be publicly observable objects of a sort that different observers can literally see, or hear, or touch. Other sorts of particulars are not basic in this respect. For example, private experiences or sense-data are identifiable only with respect to the person whose experiences or sense-data they are; and elementary particles can be identified only by reference to the macroscopic objects of which they are supposed to be constituents.[21]

Strawson's argument makes explicit the sort of grounds that underlies Bohr's thesis of the necessity to describe the instrument of observation in ordinary-language terms. The argument helps to explain what he has in mind when he says:

> Just the requirement that it be possible to communicate experimental findings in an unambiguous manner implies that the experimental arrangement and the results of the observation must be expressed in the common language adapted to our orientation in the environment.[22]

Strawson's argument, moreover, explains why Bohr maintains that continuity is a presupposition of our ordinary common-sense conceptual scheme, of the meanings of ordinary words, as Bohr puts it. It is, Strawson argues, a necessary condition of our conception of a single spatio-temporal reference frame that the material objects which constitute that frame are taken to exist *continuously* through changes of place and time. If they were not so taken, we would have the idea of a new, a different spatial frame for each new continuous stretch of observation. The idea of a re-identifiable particular implies the idea of a particular which exists continuously through temporal intervals when it is not being observed.[23] As Bohr puts it:

> . . . the definition of every concept or rather word essentially presupposes the continuity of phenomena and becomes ambiguous as soon as this presupposition no longer applies.[24]

There is a further point of affinity between Bohr and Strawson. The status of material bodies as basic particulars, Strawson argues, is a necessary condition of knowledge of objective particulars, i.e. particulars which a thinker can distinguish as objects distinct from his thought about, or experience of, them.[25] The fact that material objects are basic particulars is a necessary condition of our possession of the subject-object distinction in the way that we do, a distinction that is basic to the very concept of objective experience. This is a necessary condition, since in order for us to have objective experience, to be able to distinguish between the object of our experience and our experience of it, the object must be thought of as existing continuously through an interval during which it is not being observed. A clear distinction between the subject of experience and the object of experience is for Strawson, as for Bohr, a prerequisite of objective experience and objective discourse.[26] When this distinction cannot be clearly made, Bohr remarks, our 'forms of perception' fail, and we reach the limit of our 'capacity to create concepts'.[27] Both Strawson and Bohr share the same concern with the preconditions of objective discourse, a concern which, Strawson recognises, is reminiscent of one of Kant's main preoccupations.[28]

Strawson's argument also provides a philosophical justification for Bohr's thesis of the indispensability of the classical, ordinary-language concepts. The concepts in terms of which we describe basic particulars – macroscopic material bodies – are indispensable in the sense that the describability of such particulars is a necessary condition of our being able to communicate with others through language. Present common-sense concepts are perhaps not the only possible ones in terms of which basic particulars could be described; but they are peculiarly appropriate given the nature of our sensory equipment, and they cannot be jettisoned all at once. The common-sense conceptual scheme, Strawson argues, so thoroughly permeates our ordinary sensory experience that the character of that experience, as we ordinarily enjoy it as mature adults, cannot be faithfully described without employing ordinary common-sense words such as 'grass', 'green' etc. This is the point that Bohr has in mind when he says that 'every word in the language refers to our ordinary perception'.[29] From the fact that our ordinary perceptual experience is through and through permeated by the common-sense scheme, it does not follow, Strawson continues, that the common-sense view of the world must be true, but it does follow that unreflectively, before philosophy sets in, we unquestioningly take it to be true. It follows also that it is quite wrong to regard the common-sense view of the world as having, for the ordinary person, the status of a *theory* with respect to a type of sensory experience which constitutes the *data*, and provides the *evidence*, for it; it is wrong, because a set of beliefs is correctly described as a theory only if the data with respect to which it is putatively a theory are describable in terms that do not presuppose acceptance of those beliefs. This is not to say that we may not abstract ourselves from our ordinary, unreflective situation as perceivers, and, as philosophers, question the common-sense view, treating it, quite artificially, as if it were a theory. Addressing the question of the truth of the common-sense view, Strawson, somewhat gingerly, argues that it may be regarded as being, on the whole, true, not absolutely but relatively to our point of view as ordinary perceivers; the scientific view of the world may also be considered to be, in many respects at least, true, though again not absolutely, but only relatively to a certain conceptual point of view. Truth, for Strawson, is relative, not absolute.[30] Strawson does not propose that the common-sense conceptual scheme is the only possible one, or that the common-sense view of the world is true of reality in any absolute sense, but merely that, as things are, this scheme colours our perception of, and thought about, reality so deeply and pervasively that any radical questioning of it is difficult, if not incoherent. His metaphysics purports to be descriptive, not evaluative or prescriptive. Here again, Strawson is very close to Bohr, who, I have argued, rejects the absolute conception of truth.

It is clear that Bohr would have found Strawson's philosophy congenial. I do not wish to suggest, however, that he would have endorsed it in all details; my intention rather is to indicate the sort of sophisticated philosophical elaboration and defence of which Bohr's ideas are susceptible.

10.4
A weaker form of realism

Bohr's philosophy of physics is not scientific realism. Is it, then, some weaker form of realism, such as that conveyed by propositions (1) – (4) of Section 10.1? It is difficult to give a simple answer to this question, since quantum physics comprises many different sorts of sentences. A distinction should be made between quantum mechanics *qua* mathematical theory of motion, which may be applied both to microphysical and to macrophysical objects, and quantum physics as a general theory of matter and radiation which includes quantum mechanics, and which postulates the existence of microphysical entities such as atoms, electrons, photons, etc.

Apropos quantum physics in general, Bohr took seriously the hypothesis of the existence of microphysical entities such as elementary particles and electromagnetic waves. In 1929 for example he states that we know with certainty that some phenomena are caused by the action of a single atom, and that we have gained knowledge of the structure of the atom, indeed 'every doubt regarding the reality of atoms has been removed'.[31] He qualifies this statement, however, by remarking on 'the limitation of our forms of perception'. The point of this qualification is that we have no wholly adequate visualisable model of the atom. Bohr's construal of models in atomic physics is weakly realist in that he does attribute *some* realistic significance to the standard models of matter and radiation – the corpuscular model of the structure of matter and the undulatory model of the structure of radiation. Although we have no wholly adequate visualisable picture of them, elementary particles are more like classical particles than like classical waves, and electromagnetic radiation has a structure that is more like that of a classical wave field than that of a collection of particles. Concerning the quantum theory of matter and radiation, then, Bohr is a realist to the extent that he takes seriously the question of the realistic significance of notions like atom, electron, etc.: he accepts propositions (1) and (2) in my characterisation of realism. Whether he also accepts propositions (3) and (4) is a question I shall postpone for the moment.

What was Bohr's attitude to quantum mechanics, as distinct from what I have called quantum physics in general? It is impossible to ignore the many statements of Bohr's which invite a non-realist, instrumentalist interpretation. For example, commenting on Einstein's objections against the uncertainty principle, he writes:

> In my opinion, there could be no other way to deem a logically consistent mathematical formalism as inadequate than by demonstrating the departure of its consequences from experience or by proving that its predictions did not exhaust the possibilities of observation.[32]

It would be wrong, however, to construe such a statement as expressing an instrumentalist view of quantum mechanics *tout court*; for we ought to distinguish between different classes of quantum-mechanical sentences. Bohr's statement above refers, I believe, to the nomic or law-like sentences of the form 'the probability is r that the observable A of an object in state ψ will yield on measurement the value a'. It is probable also that Bohr intended to construe the state vector in non-realist terms, regarding it not so much as providing a description of the physical state of an individual object as a mathematical device that generates the correct probability functions for the results of measurement. He says that the wave function is a purely symbolic device which specifies the statistical laws governing observations obtainable under specified conditions.[33] There is of course no inconsistency in adopting a realist construal of the notion of elementary particles and a non-realist construal of the state vector and the nomic sentences of quantum mechanics.

But what of the experimental sentences of quantum mechanics, such as 'observable A of object O has the value a'? This sort of sentence is of prime importance for the question of a realist interpretation of the theory, since it is these sentences which above all are used to say something specific about the objects to which the theory is applied. Does Bohr construe such sentences in realist terms? Does he, that is, accept proposition (1) with respect to them?

Indeed, he does. In the clause 'certain sorts of sentences belonging to the theory may be genuinely propositional' of proposition (1), the words 'may be' are important. If 'are' were substituted for these, Bohr would not accept proposition (1) without further qualification; for, in that case, proposition (1) would be compatible with the strong form of realism that is entailed by the intrinsic-properties theory, to the effect, namely, that each experimental proposition of the form 'observable A of object O has the value a at time t' is determinately true or false independently of what human beings know or perceive. Bohr rejects the intrinsic-properties theory, and hence the strong form of realism which it entails: unlike Einstein, he holds that experimental statements are not determinately true or false *tout court*, but only under certain conditions.

It might seem that if measurement is one of these conditions, then Bohr cannot legitimately accept proposition (1), on the grounds that, on this condition, the truth or falsity of an experimental statement is not independent of human knowledge. There is, however, no incompatibility here, since an observable may be measured without our knowing the result of the measurement. Measurement is one of the meaning conditions for experimental statements only with respect to the physical operation which it involves, and not with respect to the act of cognitive awareness that normally accompanies it.

For Bohr, then, the truth-value of an experimental statement, when it has one, is independent of our knowledge and perception: but it is not also independent of our ability to know or means of knowing its truth-value, since our means of knowing are simply the experimental arrangement and the measurement procedure which define the preconditions of the meaningful assertibility of the sentence in question; and the truth-value is not independent of what we *do*.

Whether or not we say that Bohr is a realist depends of course upon our precise conception of realism. According to the widely discussed conception that has been proposed by Dummett, Bohr is not a realist. In Dummett's view, the primary tenet of realism, with respect to a certain class of statements, is that each statement of the class has a determinate truth-value, whether or not we know or are able to discover it, the truth-value being determined by some objective reality whose existence and constitution are independent of our knowledge of, or capacity to know, it.[34] It is a necessary condition, but not a sufficient condition, of one's adopting a realist interpretation of some class of statements that one regards the law of bivalence – to the effect that each statement of the class is either true or false – as applying without restriction.[35] Bohr can be read as denying that the principle of bivalence holds for the class of experimental statements *tout court*: these statements have truth-value only under certain conditions. In this respect his interpretation is non-realist, given Dummett's conception of realism. Dummett recognises, however, that the sort of non-realism (or 'anti-realism' as Dummett calls it) which results from restricting the scope of the principle of bivalence may be very weak.[36]

Dummett's conception of realism is much stronger than the one which I expressed in proposition (1); it is in fact the strong sort of realism that Einstein favours. Contrasting Bohr's position with Einstein's, it is fair to say that Bohr is not a realist. Bohr, however, is rejecting only strong realism, not realism *per se*; and if we contrast his stance with a strong anti-realism, it is fair to call him a weak realist, for there is a weaker, and indeed subtler, realism that he is prepared to accept. It is for this reason that I prefer to say that his position is weakly realist. Nevertheless, as I shall argue in the following sections, the weak realism which I am attributing to Bohr will require further qualification, indeed further weakening.

10.5
The mathematical structure of physical reality

Bohr holds that the experimental statements of quantum mechanics, under appropriate conditions, have determinate truth-values, and hence describe physical reality. But how, exactly, does he conceive of the properties which these statements predicate? These properties are inherently quantitative. How, then, is their mathematical character, and the mathematical structure which they presuppose, to be understood? In Bohr's view, does physical reality have a structure that is faithfully describable in terms of some mathematical theory, a structure which we come to discern in it rather than one which we imaginatively read into it?

The key to understanding Bohr's position here lies, I believe, in his attitude to the question of the relation between pure mathematics and the real physical world. Whether or not he held a Platonist or a constructivist view of the cognitive status of the propositions of pure mathematics, he held a non-realist view of the cognitive status of applied mathematics – non-realist in the sense that he was highly doubtful of the view that the physical world has a structure that is uniquely describable in terms of some theory of pure mathematics, it being the task of physicists to discover what this correct mathematical theory is. In Bohr's opinion, a theory of pure mathematics is an ideal construction of reason, the objects of which (if any) are the entities of some abstract, ideal realm: so far as its applicability to the real physical world is concerned, its value is primarily instrumental; it is simply a useful tool for aiding our understanding of our experience of the world, though it may not be entirely fictitious, capturing nothing of reality as it is in itself.

It is useful to compare Bohr with Einstein on this matter. Bohr's view of the nature of theoretical concepts is in some respects very similar to Einstein's: both men hold that physical concepts are constructions of the mind; they are neither given in experience nor *a priori*. This is what Bohr has in mind when he states that 'from our present standpoint, physics is to be regarded not so much as the study of something *a priori* given, but rather as the development of methods for ordering and surveying human experience'.[37] Bohr does not hold, however, that theoretical concepts are arbitrary constructions or 'free inventions'; these concepts, rather, have their roots in everyday physical notions, of which they are rational elaborations. Physics is a refinement of the common-sense conception of the material world as a system of materially interacting, localisable entities capable of locomotion within a spatio-temporal framework. The classical physical concepts are entrenched in our thought not because they are *a priori*, but because of their relative proximity to our ordinary common-sense conception of the world.[38] Unlike Bohr, Einstein was inclined towards a Platonist conception of applied mathematics – the 'Mathematics is the language of the Creator' view – according to which physical reality has a structure that may be faithfully described by some mathematical theory. For Bohr, mathematics is merely the indispensable handmaiden of physics. He frequently speaks of irrational and complex numbers as 'conventional devices': quantum mechanics and relativity theory 'rest essentially on the old mathematical artifice of the introduction of imaginary quantities', the quantum of action and the velocity of light for example being factors $\sqrt{(-1)}$.[39] This attitude explains why he was never greatly interested in the mathematical formalism of a physical theory as such; his primary concern was the physical interpretation of the formalism, the question of what qualities were to be associated with the quantities of the theory.

For Bohr the notion of a moving object's having an exact simultaneous position and momentum is a logically coherent concept, but it does not necessarily have any realistic application in quantum physics – or classical physics for that matter. Applied to macroscopic objects such as billiard balls, the notion was useful in that it facilitated the construction of a theory which had great explanatory and predictive power. The failure of this notion in quantum physics however, serves to remind us that the concept is an idealisation which *may* characterise physical reality, but which need not and, it seems, does not. Strictly speaking nothing falls under the notion – not even macroscopic bodies such as billiard balls – and hence there are no such things as particles in the classical sense.

Einstein disagrees with Bohr on this issue. In his view, macroscopic objects, such as billiard balls, have at all times an exact, or at least approximately exact, simultaneous position and momentum; yet, if a billiard ball is described in quantum-mechanical terms, there are circumstances in which the ball cannot be said to have even an approximately exact position. In a dispute with Born, he argues that, for a macroscopic ball rebounding between two walls, quantum mechanics predicts an enormous indefiniteness in its position over a long period of time; yet it would be absurd to suggest that the act of observation creates the quasi-exact position which we find when we look at the ball; hence the ball must have had a quasi-exact position all along.[40]

Commenting on this dispute, Pauli rejects Einstein's conclusion. In Pauli's opinion, it is false that a macroscopic body always has a quasi-exact position, since there is no fundamental difference between microphysical and macrophysical objects in this respect. Observation of the ball, then, does indeed 'create' the observed position, though the observation is 'a "creation" existing outside the laws of nature'.[ib.] Pauli's point here is that the 'creation' of the position is not a physical process at all; but rather, the quasi-exact position is brought about, so to speak, by the establishment of the experimental conditions that are required for the observation; these conditions constitute the meaning conditions for the predicate in question. Pauli remarks that Einstein has the philosophical prejudice that for macroscopic bodies 'a state (termed "real") can be defined "objectively" under *any* circumstances'.[41] Pauli's view is clearly the same as Bohr's.

Einstein's view is in fact more subtle than Pauli suggests. He holds that the notion of the real state of an object is not given *a priori*, but rather is defined 'within the framework of the chosen conceptual system'.[42] Thus it is not *a priori* that a macroscopic object such as a billiard ball has at every instant a quasi-exact position; but within the conceptual framework of classical physics that is how an object is to be conceived, and since quantum mechanics retains the conceptual apparatus of the continuum of real numbers, it provides no compelling reason to reject the notion of quasi-exact simultaneous position and momentum. The conceptual systems of classical mechanics and quantum mechanics have the *same* framework in this respect. Einstein is not the naive realist that Pauli takes him to be. How could he be, given the avowed Kantian slant to his thinking which we observed in the previous chapter?

While Einstein agrees with Bohr that it is not *a priori*, or necessarily true, that the real number system faithfully describes the properties of physical reality, he differs from Bohr in wishing to retain a realistic construal of that system in quantum mechanics. Bohr, by contrast, holds that although the mathematical frameworks of the two theories have much in common, there are important differences in their conceptual structures, and hence what is construed realistically in classical mechanics need not be so construed in quantum mechanics.

Some may be attracted to a more naive view than Einstein's. It may be argued that physical space and time have a structure that is similar to that of the real number system in certain respects. Space may be thought of as consisting of a non-denumerably infinite set of spatial points, and time a non-denumrably infinite set of temporal instants, the points and instants being dimensionless limits rather than intervals. Since there is a one-to-one correspondence between points and instants and the real numbers, an object, whether moving or motionless, is at some point at each instant. To deny this is to be committed to the notion of an object's having a fuzzy or ontically inexact position. But that notion, the argument continues, is incoherent: at each instant the point which constitutes an object's centre of mass must be congruent with some unique point in three-dimensional space. There can be vague predicates, whose meaning is not defined in all respects, but no vague or ontically indefinite properties; vagueness is a semantic or conceptual, rather than ontic, characteristic.

This argument, as it stands, is unpersuasive, since the view that vagueness is exclusively a semantic property is no longer generally taken for granted;

and it is by no means self-evident that real space and time have structures that are truly describable by the real number system; indeed, it is a moot metaphysical point. Points and instants thus conceived are abstract, ideal items which it is not only practically impossible, but also logically impossible, to perceive with the senses; and even if we construe them as the limits of converging series of nested intervals (by A.N. Whitehead's method of extensive abstraction, say) it is arguable that they should be given any realistic interpretation at all. Moreover, as Teller has argued, the precise mathematical notion of exact position in classical physics is a refinement of the intrinsically vague, prescientific notion of position, the notion of a vague location within a certain spatial region. The classical notion has become so deeply entrenched in our thinking that we are inclined to think that corpuscular objects must have absolutely exact positions. But the classical notion is not the only possible mathematical refinement of the ordinary conception of position; it is not *a priori*. Moreover, it is a notion that has outgrown its usefulness, since quantum mechanics, while preserving the scheme of the real number continuum, has no proper use for the concept of absolutely exact position; it operates with a notion of position as something essentially inexact. Whereas in classical mechanics an observable may in theory have a value that is absolutely exact, in quantum mechanics observables which are continuous quantities, such as position and momentum, have no absolutely exact values, since they have no eigenvalues which are point-values; indeed, strictly speaking, they have no eigenvalues at all, though quasi-eigenvalues – intervals of point-values – are constructed for them.[43]

Unlike Einstein, Bohr was sympathetic to a non-realist construal of the application of pure mathematics to the description of real space and time, and in particular to the representation of their structure in terms of the real number system. On a weakly non-realist construal of the notions of point-location and point-instant, one would be agnostic as to the question whether these notions denote real items; and on a more strongly non-realist view, one would deny that they had any realistic significance at all; these notions would be conceived of merely as devices which facilitate the formation of an empirically adequate mathematical theory of motion. Bohr's view here, I believe, was non-realistic, at least in the weak sense – and perhaps even in the strong sense.

Not only is Bohr disinclined to give the predicate 'has an exact position and an exact momentum at time *t*' a realist interpretation, but he is inclined even to doubt whether the predicates 'has an exact position at t' and 'has an exact momentum at t' should be construed in strongly realist terms. The properties which these predicates express are, he holds, mathematical abstractions which *may* truly characterise physical reality, but which need not. Physical reality cannot be presumed to be in itself exactly as it is described by the mathematical concepts. This is not to say that Bohr is rejecting the notions of truth and falsity as applied to such predicates – if he did, he would in no sense be a realist; it is merely to say that he rejects a notion of truth and falsity as being absolute properties of statements. An experimental statement, if true, is true not absolutely, but only relatively to the conceptual point of view which we human thinkers adopt. Bohr's conception of truth is similar to the relativist notion which is presupposed by Strawson's descriptive metaphysics. Bohr, then, is not rejecting the notion of truth, but merely the notion that the truth of a thought consists in its being a faithful conceptual picture of reality as it is in itself. Bohr's acceptance of proposition (1), Section 10.1, therefore, must be qualified by pointing out that he does not have an absolutist conception of truth.

This qualification has consequences for the question of whether Bohr accepts proposition (3) – 'the main aim of physics is the construction of theories which approximate as closely as possible to the truth about physical reality'. Bohr would have been chary about accepting this proposition as it stands. Here is where the main difference between Bohr and Einstein lies. Einstein holds that the primary aim of physics is to comprehend physical reality. Yet we cannot do this simply by reading off the intrinsic properties of reality from our experience; nor are these properties given to us *a priori* by human reason. Our concepts of such properties have to be constructed. Yet, Einstein believes, some of the mathematical concepts which we freely invent may in fact provide true representations of the intrinsic properties of physical reality. The fact that many of the concepts which we invent seem to fit reality as well as they do is for Einstein something awesomely mysterious. For Bohr, however, the primary aim of physics is to help us make sense of our perceptual experience. He does not doubt that our experience is of an independently existing physical world, but he denies that the main aim of physics is the comprehension of the imperceptible structure of that world for its own sake. We construct theories of the microstructure of the physical world in order better to comprehend the world in which we move and have our being – the macroscopic world.

10.6

Bohr: an instrumentalistic realist

Is Bohr, after all, a weak instrumentalist of the van Fraassen sort, rather than a realist of any description? Some of his statements might suggest that he is; for example:

> ... in our description of nature the purpose is not to disclose the real essence of the phenomena but only to track down, so far as it is possible, relations between the manifold aspects of our experience.[44]

Nevertheless, it would be quite wrong to describe Bohr as a weak instrumentalist, because for the latter the truth, as distinct from the empirical adequacy, of a physical theory is of no concern whatever. Now, while holding that the primary aim of physics relates to experience rather than to reality, Bohr is in fact interested in the question of whether a physical theory is true. He does after all accept propositions (1) and (2) Section 10.1 with appropriate qualifications; and again, with suitable qualifications, he would accept proposition (4). There is simply no denying these weakly realist components in his philosophy of physics.

Proposition (3) is the main stumbling block to his accepting the reasonably sanguine realism that is defined by propositions (1) – (4). Because he regards physical theories *primarily* as tools for the conceptual organisation of our sensory experience, it is fair to recognise a certain weakly instrumentalist ingredient in his thinking, an ingredient which is even weaker than van Fraassen's weak instrumentalism. I propose, therefore, to describe his philosophy as *instrumentalistic realism*, at the expense no doubt of an air of paradox: the realist component and the instrumentalist component are, so to speak, complementary sides of the phenomenon that is Bohr.

10.7

The philosophical grounds of the indefinability thesis

It is time now to return to the question that was shelved at the end of Section 7.7 viz. the question concerning Bohr's grounds for maintaining the strong meaning condition, to the effect that the making of a measurement is a necessary precondition of the meaningful applicability of a physical predicate. What is his theory of meaning for such predicates?

Bohr's theory is not realist in the sense that the meaning of a predicate or an experimental statement is taken to consist solely in its truth conditions, i.e. the states of affairs which, were they to obtain, would render the predication or the statement true, these states of affairs being independent of our knowledge of, or ability to know, them. There is a non-realist component in the meaning of a predicate or experimental statement in that it is a necessary condition of the meaningful assertibility of such a statement that conditions obtain which would enable us to determine its truth-value.

Bohr's theory has something in common with the non-realist theory of meaning which has been sympathetically discussed by Dummett. According to this theory, the meaning of a statement is constituted not by its truth conditions, but by its verifiability or assertibility conditions, i.e. the conditions which would enable us to discover its truth-value or which would rationally justify our asserting it. In order to know the meaning of a statement, what one must know are its verifiability or assertibility conditions. Dummett gives the following argument for this thesis.

Language is a tool which is used for the purposes of communication. What understanding the meaning of a linguistic expression consists in is the practical knowledge of how to use it for the purposes of communication. To understand the meaning of an indicative sentence, for example, is to know how to use it to make a specific descriptive statement. Owing to the intersubjective nature of communication, this practical knowledge must be publicly displayed in the language user's linguistic behaviour, in his uses of sentences and in his responses to other speakers' uses of them. It is through the public display of this practical knowledge that the meanings of words and sentences are learnt. If, however, the meaning of a sentence consists simply in its truth conditions, and knowing its meaning consists in knowing its truth conditions, then, in the case of sentences whose truth conditions are such that it is impossible for us ever to recognise whether or not they obtain (such as mathematical sentences which quantify over infinite domains), this practical knowledge cannot be manifested in the use of such sentences. What we display in using them can be nothing more than our grasp of their assertibility conditions. If the truth conditions of a sentence are such that it is impossible for us to recognise whether or not they obtain, then it makes no sense to say that someone has implicit knowledge of these conditions, since there is no practical ability by means of which such knowledge could be manifested.[45] On this view the meaning of a statement cannot intelligibly be held to transcend our capacity to ascertain its truth-value, or to transcend all humanly accessible grounds for asserting it.

It is likely that Bohr would have found this argument congenial, since he holds that the meaning of an experimental statement consists, partly at least, in its verifiability or assertibility conditions, these being the presence of an appropriate measuring instrument and the occurrence of an appropriate measurement. Dummett's argument illustrates the point that a sophisticated defence of the non-realist component in Bohr's theory of meaning is available, a defence, moreover, which is wholly independent of empiricist or positivist foundations. The logical positivists are not the only philosophers to have adopted a verificationist theory of meaning: the intuitionists in the philosophy of mathematics hold a verificationist theory, the basis of which is not empiricism; indeed, they are radically opposed to an empiricist philosophy of mathematics. And of course Dummett's views have been inspired by intuitionism, not logical positivism.[46]

Given the verificationist or confirmationist component in Bohr's theory of meaning, his realism requires further qualification. He can consistently accept the conception of realism that is expressed in proposition (1) Section 10.1, but the notion of truth which is involved in that conception must itself be construed in weakly realist terms. The notion of truth must contain a reference, if not to human knowledge, at least to the possibility of acquiring knowledge. On this view of truth, a statement is true only if it is possible for us to discover the grounds which would either verify it or confirm it. Bohr of course is not required to say that truth consists simply in verification or confirmation, but merely that a reference to verifiability or confirmability is part of the very meaning of the concept.

What, then, is the philosophical basis of the verificationist element in Bohr's theory of meaning, if it is not empiricism? It is, I believe, pragmatism. Pragmatism in fact began as a theory of meaning. According to Peirce, the father of pragmatism, a physical predicate has a well-defined meaning only if it can be used to make an assertion which has effects that might conceivably have a bearing on practice, and the meaning consists largely in the sum of these effects.[47] Since one of the most prominent practical effects of an assertion concerns the actions which are involved in verifying it or discovering grounds for believing it, there is a distinct verificationist element in the pragmatist theory of meaning.

The basis of Bohr's theory of meaning is pragmatism: the ascription of an exact position and an exact momentum to an object at the same time is meaningless, not principally because such a property is unobservable, but because the ascription has no practical consequences whatever: it has no explanatory or predicative power; it 'does no work', as a pragmatist would put it. The fact that the statement is not verifiable is only one factor among others that makes it meaningless to assert it. The meaning conditions, then, are not primarily epistemic but pragmatic, having more to do with what we as agents can *do* than with what we can *know*. Nevertheless, for Bohr the verifiability or confirmability of a statement is a necessary condition of its meaningful assertibility; and since experimental statements are verifiable or confirmable only by means of some physical measurement operation, his theory has an operationalist component. The operationalist component is already present in Peirce's theory of meaning. He holds that the cognitive meaning of a predicate is exhaustively defined by specifying all the experimental phenomena which the affirmation or denial of it could imply: for example, what it means to say that something is hard, is to say that it resists scratching, and the statement '*A* is hard' is meaningful only if it is possible to scratch *A* and observe the results.[48]

Pragmatism is also the basis of the weakly instrumentalist component in Bohr's philosophy which we were compelled to acknowledge: it is at the root of his reluctance to accept proposition (3). The primary aim of physics, in Bohr's view, is to make sense of our perceptual experience, and the description of microphysical reality is merely a means to that end, not an end in itself; and physical theories primarily are useful tools or instruments to the fulfilment of that end. It is pragmatism, moreover, that is the ultimate basis for Bohr's indefinability thesis.

10.8
Høffding and the historical roots of Bohr's pragmatism

What historical evidence is there for my assertion that the philosophical foundation of the indefinability thesis is pragmatism? Bohr's general philosophy has its roots in that late nineteenth-century tradition which tried to steer a middle course between the extremes of realism and idealism, a tradition which took slightly different directions in neo-Kantianism and pragmatism. Hans Vaihinger and William James respectively are good representatives of that tradition. Both philosophers were agreed that the adequacy of a concept construed as a representation of reality depends not upon the extent to which it faithfully 'mirrors' reality (since, they held, reality is not graspable independently of the medium of a conceptual framework) but upon its expediency, upon its success in aiding a comprehension and extension of our perceptual experience. Bohr's conception of physical concepts as abstractions and idealisations reminds one to some extent of Vaihinger's thesis that theoretical concepts generally are fictions, in the sense of being artifices, products of the rational imagination which cannot be taken to be 'true reflections' of reality; indeed Vaihinger regards many theoretical concepts as self-contradictions (as for example the notion of mass-point) which are nevertheless of instrumental value – they are 'instruments for finding our way about more easily in this world'.[49] We conceive of reality *as if* it were as our concepts portray it. This form of instrumentalism is, however, stronger than the variety which I have ascribed to Bohr. Much closer to Bohr than either Vaihinger or James is Høffding, an old and close friend of the Bohr family. Høffding is the source of the pragmatist element in Bohr's philosophy.

Høffding's philosophy has much in common with that of James. The two philosophers were exact contemporaries, and admirers of each others' work. Høffding paid a visit to James at Harvard in 1904. Høffding's philosophy is succinctly set out in his *The Problems of Philosophy*, which was translated into English at the request of James.

Høffding sees the main task of epistemology as being to uncover the basic presuppositions upon which our understanding of our experience is based – a conception of epistemology that, as Høffding acknowledges, originates with Kant.[50] One is struck immediately by the similarity of this tenet to one of Bohr's philosophical preoccupations. Where Kant went wrong, in Høffding's opinion, was in thinking that the fundamental presuppositions can be specified once and for all; he was right, however, in thinking that the demand for unity and continuity is fundamental to our understanding of our experiences. All the Kantian categories can be traced back to the concept of continuity: it is this notion that underlies the principles of logic and science:

> It is the ideal of knowledge to find it in all domains of observation. Our mind can only understand by synthesis, and the principle of continuity is therefore the presupposition, the working hypothesis, of all science.[51]

The relation between continuity and discontinuity, Høffding maintains, lies at the root of all the major philosophical problems; there is in fact a fundamental duality between continuity and discontinuity.[52] Again, one is strongly reminded of Bohr here.

Høffding rejects the correspondence theory of truth, on the grounds that there can be no comparison between thought and things, but only between thought and experience. He adopts a pragmatist theory of truth according to which the truth of a principle consists in its 'working value', in its 'firmly ordering and unifying phenomena': principles are working hypotheses in that, quite literally, they *do work*. He writes, for example:

> It is not necessary that principles have other qualities than those which are required to the obtaining of this aim. It is not necessary that they contain an absolute reproduction of the innermost essence of Existence. Their aim is to make an order of phenomena possible, and this they can do, even if they contain figurative expressions, if only . . . the connexion of phenomena can be expressed and foretold by it.[53]

The truth of a principle does not consist in its 'conformity to an absolute order of things'; indeed, we ourselves produce truth when we find the principles which connect the greatest possible number of phenomena.(ib.) Høffding calls this the *dynamic*, in contrast to the *static*, concept of truth – dynamic, since the truth of a statement consists in its energy as it were, in its capacity for work, and also since the truth is not something fixed, but something that changes as our ideas evolve. He also calls it a *symbolic* concept, since truth indicates not an absolute, but a relative, similarity or analogy between thought and its real object.[54] Concepts are symbols of reality by means of which we order our experiences. Moreover, different sets of concepts may order experience equally well: no single conceptual scheme can be said to be the only possible one or the best. A conceptual scheme is adequate if it is consistent and capable of representing new phenomena.[55] This view is similar in some ways to James's doctrine that theories are 'instruments', 'mental modes of adaptation to reality' and that truth is the expedient in the way of thinking – the expedient on the whole and in the long run. James regards the notion of reality independent of particular experiences as an instrumental, 'regulative' idea; he rejects both positivism and naive realism.[56] Høffding's instrumentalism, however, is not as strong as James's: Høffding maintains that reality itself is to some extent intelligible – to a certain extent we can penetrate into reality itself.[57] This is exactly the attitude that Bohr takes.

Høffding's philosophy of physics follows similar pragmatist lines. The concepts of physics provide primarily an economical organisation of experience. The success of the mechanical world view is due to the mathematical theory of continuity with which it operates; but there are no grounds for thinking that the quantitative properties expressed by physical concepts 'express the inner-most essences of things': the concepts of

mathematical physics are merely useful images or symbols.[58] The principle
of causality, for example, cannot be presumed to have universal validity.[59]
There is, besides, an *irrational element* in the relation between our thought
and reality, between quality and quantity, between subject and object: 'being
may possess attributes that cannot be comprehended or defined by means of
dimensions in which our thoughts can move':[60] there are limits to the degree
of continuity that thought can establish among experiences.[61] Høffding
speaks of 'this incommensurability, or this (in a mathematical sense)
irrational relation, between thought and reality'.[62] In holding that complete
continuity cannot be attained, that no single conceptual scheme is adequate
for the comprehension of the whole reality, Høffding was following Søren
Kierkegaard, who maintained that reality is so full of difference and
opposition that no single system of thought can adequately represent it.[63] It
is this idea that lies behind Bohr's notion of wave-particle complementarity.
Bohr in his youth was, like Høffding, deeply impressed by Kierkegaard's
writings, though, like Høffding, he was by no means in complete agreement
with Kierkegaard.[64] Owing to Kierkegaard's influence, however, the ideal of
conceptual homogeneity never appealed to Bohr, as it did to Einstein, as a
fundamental goal of physical theory.

The following passage brings out the pragmatist character of Høffding's
philosophy of science so well that it is worth quoting at length:

> It is now more and more admitted that the importance of scientific and
> philosophical principles consists in this, that they lead us in our striving after
> understanding. Their truth is their validity, and their validity is experienced in
> their power of leading us in our intellectual work. A principle is true if it can be
> applied, if we can work with it, *i.e.*, gain understanding with the help of it.
> Truth is a dynamic concept; it manifests itself in the working of our thought.
> And it is a symbolic concept, because it only presupposes an analogy, not an
> identity between thoughts and events. This holds of the truth of our sense-
> qualities; they have objective value as symbols, but can not be proved to be
> images of things. It holds, too, of the truth of our formal-logical principles, of
> the principle of causality, etc. We cannot compare our sensations and our
> principles with an absolute order of things. Surely we have no right to regard it
> as a pure accident that just these special sensations and thoughts make it
> possible to gain a progressing knowledge of the events of the world; but neither
> have we the right to regard them as direct revelations.[65]

Here we see, clearly and distinctly, the origin of the weak instrumentalist
component in Bohr's philosophy.

The pragmatist strain in Bohr's thought was first recognised by H.P. Stapp. Stapp, however, does not observe that the source of this strain is Høffding, not James.[66] The influence of James on Bohr's thinking was first suggested by K.M. Meyer-Abich; but it is James's psychological writings, rather than his philosophical works, which Meyer-Abich thinks were influential, and he makes only one passing reference to James's pragmatism.[67] In an interview with Kuhn and Aage Petersen which he gave just before he died in 1962, Bohr states that he read James before going to Manchester in 1913.[68] It is very likely that what he read then were some of James's philosophical writings recommended to him by Høffding. I shall not pursue this historical point, however, for I am convinced that the source of Bohr's pragmatism was Høffding himself, and not James. The influence of Høffding has been discussed by Gerald Holton and L.S. Feuer; but Holton is vague on the exact nature of the philosophical influence, and Feuer stresses – overstresses in my view – the Kierkegaardian side of Høffding.[69]

10.9
The Kantian elements in Bohr's philosophy

Høffding is the source of the Kantian overtones of some of Bohr's statements. There is no evidence to suggest that Bohr had any first-hand acquaintance with Kant's writings. Høffding, though not himself a Kantian, was something of a Kant scholar, and sympathetic to neo-Kantianism. As I have already remarked, Bohr's use of the Kantian phrase 'forms of perception' is distinctly Høffdingian.[70]

Some commentators, most notably Hooker, have stressed the Kantian overtones, and John Honner, elaborating on Hooker's reading, has gone so far as to label Bohr's philosophy 'transcendental', meaning by this that Bohr is concerned primarily with the necessary conditions of the possibility of objective knowledge.[71] The label 'transcendental' is in some respects apt, for Bohr was concerned with the conditions for the possibility of objective knowledge of microphysical objects. These conditions, he held, are that there should be a clear distinction between the object and the instrument used to investigate it, and that the instrument should be describable in classical terms.

What exactly is the putative analogy between Bohr's philosophy and Kant's? For Bohr, ordinary-language concepts have a role to play which is reminiscent of the role which the categories play in Kant's philosophy. Moreover, Bohr's distinction between the object *per se* and the object as it appears in the context of a measurement (the 'phenomenon') reminds one of Kant's distinction between the object *qua* thing-in-itself (the *noumenon*) and the object *qua* thing-as-it-appears (the *phenomenon*). Kant and Bohr both hold that the object is an object of meaningful discourse and possible knowledge only *qua phenomenon* (in the different respective senses which they give this term). There is something illuminating in this analogy. In the chapter of *The Critique of Pure Reason* entitled 'The Ground of the Distinction of all Objects in general into Phenomena and Noumena', Kant argues that concepts are meaningfully applicable only to objects of possible sensory experience – to phenomena. Kant is concerned primarily with *a priori* concepts, but the argument is intended to apply to all concepts. His argument is that a concept is meaningful only if we can 'give it an object to which it may be applied'; but the only objects that are accessible to us are objects of possible sensory experience; consequently concepts are applicable only to such objects. In the absence of such an object, a concept 'has no meaning [*Sinn*] and is completely lacking in content [*Haltung*]':

> Therefore all concepts . . . relate to empirical intuitions, that is, to the data for a possible experience. Apart from this relation they have no objective validity, and in respect of their representations are a mere play of the imagination or of the understanding.[72]

Concepts, Kant maintains, are meaningfully applicable only with respect to the mode in which we are able to apprehend the objects to which they are applied: our thought cannot transcend the limits of the mode of apprehension within which objects can be given to us.[73] The conditions for the meaningful applicability of a concept to an object cannot be separated from the conditions of our becoming aware of the object. The reason why Kant ties the applicability of concepts so closely to the possibility of sense perception is that he believes that the function of concepts is to organise ('synthesise') the multifarious data of sense perception (the 'manifold of intuition'): concepts are, as it were, instruments for transforming the data of sense into objects of experience. Concepts and objective experience are for Kant correlative: we cannot have the one without the other. This is what Kant means by his famous statement: 'Thoughts without content are empty, intuitions without concepts are blind'.[74] Kant's tacit theory of meaning has a verificationist component: to put the point anachronistically, if a concept is not applicable to an object which can be given in experience, there is no way in which it can be given any clear sense, and hence there is no way that it can have any clear reference. Strawson calls this thesis 'Kant's principle of significance', the principle to the effect that 'there can be no legitimate, or even meaningful, employment of ideas or concepts which does not relate them to empirical or experiential conditions of their application'.[75]

Analogously, Bohr holds that physical concepts such as 'exact position' and 'exact momentum' are meaningfully applicable to microphysical objects, since such objects are observable; but they are applicable *only to the extent* that such objects are observable. Now there are respects in which such objects are not observable: in conditions, for example, in which the position, say, of an object is observable, its momentum is not observable, and vice versa; under such conditions talk of the momentum of the object is devoid of cognitive content. For Bohr what matters is not that microphysical objects are too small to be observed with the unaided senses, but rather that, owing to the quantum of action, there are limits to the *extent* to which they can be observed. There are, that is, conditions under which some of the physical properties of an object have the status of being 'noumenal'; in which case the question of whether an object has or lacks, say, an exact momentum is cognitively meaningless.

While this analogy is illuminating, it ought not to be overstressed. To call Bohr's philosophy 'Kantian' or 'transcendental' *tout court* is to ignore the other important aspects of his philosophy which I have been at pains to bring out. Besides, from a historical point of view the Kant-like aspects of Bohr's thinking have their source in Høffding, and not directly in Kant. It is worth pointing out, however, that pragmatism has its ultimate source in the philosophy of Kant.

10.10
The pragmatist strain

The pragmatist strain in Bohr's thought is of crucial importance for an understanding of his philosophy of physics: it is the basis of what I have called his instrumentalistic realism. It explains why he has so often been mistaken for a positivist or radical empiricist. It explains also why there is no evidence of any major disagreement between Bohr and colleagues, such as Pauli and in earlier years Heisenberg, whose philosophy was unmistakably positivist. The positivists and the pragmatists share an abhorrence of speculative metaphysics, regarding as meaningless ontological questions that do not admit of a decisive answer on the basis of experience or human interests. Their approach to metaphysics varies from the extreme anti-realism of strong instrumentalism to the weak instrumentalism of Høffding. While the positivist and the pragmatist have much in common, there is a significant difference between them: whereas for the positivist what we can know depends upon what we can *observe*, for the pragmatist what we can know depends upon what we can *do* (which of course includes observing). The foundation of Bohr's instrumentalistic realism is pragmatism, and a pragmatism, it is important to stress, of a sort that is weaker than the very forthright doctrine of James.

Bohr's pragmatism explains also many of the details of his interpretation of quantum physics that would otherwise appear puzzling (or blatantly positivist). For example, his theory of models has both a realist and a non-realist component. The pragmatist, unlike the strong empiricist, can adopt a realist construal of hypothetical entities. On the one hand Bohr regards electrons as worthy candidates for being regarded as real entities, yet on the other hand he regards the particle model of the electron as descriptively inadequate, since it fails to cover all the empirical data. Still, he is willing to tolerate this inadequacy, employing the model in a highly restricted way, rather than attempting to fashion a more adequate model; he does so simply because the model can be made to work. He is prepared even to employ the wave model of the electron, in appropriate conditions in which the particle model is inapplicable, believing full well that the wave model has no genuine realistic significance, it being merely a picturesque metaphor of the failure of the particle model; he is prepared to do this for the sake of other theoretical benefits. Someone who holds a more realist theory of models would find Bohr's practice very unsatisfactory, for the complementary use of the dual models provides no intelligible conception of the real nature of the electron at all; indeed it makes its nature seem highly mysterious. Adopting a pragmatist approach, however, Bohr is content, since the complementary use of the disparate models is logically consistent and enables *all* empirical data to be subsumed.

Again, Bohr's pragmatism explains his ontic, or semantic, construal of kinematic-dynamic complementarity, the thesis that an object cannot be said to have an exact simultaneous position and momentum. A realist, like Einstein, may be troubled by the idea that a particle-like entity can lack an exact simultaneous position and momentum: surely an electron must be said to have such a property even though, for ineluctable reasons, we are unable to observe it or measure it. For the pragmatist, however, there is nothing *a priori* in the notion that a material object, of whatever size, has an exact simultaneous position and momentum. It is a notion that has worked well when employed within a certain range of our perceptual experience; but there is no necessity in its employment; rather, it is to be employed only to the extent that we cannot do without it, and where it can be put to no practical use, it may, and ought, to be discarded. And of course in quantum physics it has no practical use. We can now readily understand why Bohr denies that so-called 'trivial joint measurements' of the sort discussed by Park and Margenau (p.82 above) constitute adequate assertibility conditions for the predicate 'exact simultaneous position and momentum'. The ascriptive use of this predicate, on the basis of such a 'measurement', is cognitively meaningless, not just because it is unverifiable, but also because it does no work: it provides no basis for prediction or retrodiction – it is 'an abstraction, from which no unambiguous information concerning the previous or future behaviour of the individual can be obtained'.[76] For the pragmatist, what objects we treat as real and what real properties we attribute to them are questions to be settled not by some abstract, absolute conceptual scheme that is binding *a priori*, but by our human needs, interests, practices and experiences: reality is what we are constrained to postulate for the purposes of explanation and prediction. Thus the pragmatist can readily accommodate the idea of a particle-like entity's having an ontically inexact or fuzzy position and momentum at any instant. This is not to say that the realist cannot coherently admit such a notion, but is simply to say that he may find it more difficult to discard notions which are entrenched in his conceptual scheme.

Once the pragmatist foundation of Bohr's philosophy has been grasped, the Bohr–Einstein debate can be viewed from a new, and more illuminating, perspective. On the one hand, those whose sympathies are realist will be inclined to judge that Einstein was the victor. On this view, given that quantum mechanics employs the mathematics of the real number continuum and the notions of exact position and exact momentum, then there is no good reason to think that extremely small bodies such as electrons do not have these properties simultaneously; certainly our inability to observe that they do is no compelling reason to think that they lack such a conjunctive property. Moreover, the fact (if it is a fact) that we cannot put the notion of exact simultaneous position and momentum to any practical explanatory or predictive use is not a compelling reason to discard it. Besides, the notion has *some* explanatory content: it explains the otherwise puzzling fact that we can measure either the exact position of an electron or its exact momentum (but not both) whichever and whenever we choose. What better explanation of this could there be than the fact that an electron has both of these properties at all times? Furthermore, in the EPR argument, Einstein brilliantly showed that quantum mechanics itself strongly suggests that an electron has both of these properties at any moment; that at any rate is a very plausible interpretation of the results of the EPR experiment.

On the other hand, those whose leanings are more towards pragmatism or a pragmatist conception of reality will give the palm to Bohr. On this view, whether or not a predicate can be given a meaningful application depends upon the work to which it can be put. If in physics the use of a theoretical predicate has no practical effects – no substantive explanatory or predictive use – it is vacuous. Bohr holds that the notion of exact simultaneous position and momentum is vacuous in just this respect: the notions of exact position and exact momentum can be put to significant use only in the context of well-defined experimental situations. Beyond these contexts the use of such notions is a piece of speculative metaphysics, devoid of any cognitive significance. Bohr insists that experimental procedures define the contexts for the meaningful use of such predicates not so much because these procedures are required for observation or verification, but more because they provide the only contexts in which the use of such predicates has practical effects, substantive explanatory or predictive significance. For Bohr the notion of exact simultaneous position and momentum – the notion precisely of a classical particle – is a sort of conceptual appendix that has lost whatever useful function it once had.

Viewed from the pragmatist perspective, the philosophical core of Bohr's reply to Einstein's EPR argument is laid bare with greater clarity. This core is conveyed in the following enigmatic passage:

> Of course there is in a case like that just considered no question of a mechanical disturbance of the system under investigation during the last critical stage of the measuring procedure. But even at this stage there is essentially the question of *an influence on the very conditions which define the possible types of predictions regarding the future behavior of the system.* Since these conditions constitute an inherent element of the description of any phenomenon to which the term "physical reality" can properly be attached, we see that the argumentation of

the mentioned authors does not justify their conclusion that the quantum-mechanical description is essentially incomplete.[77]

The first point that is being made here is that in the EPR situation the measurement of S_1 does not, in any relevant way, physically disturb S_2; but it does preclude our making some sorts of predictions about the future behaviour of S_2. If, for example, we measure the position of S_1, then we can make a prediction about the position of S_2, but not about its momentum. One wonders, however, what Bohr is getting at when he says that the conditions defining these predictions on S_2 'constitute an inherent element of the description of any phenomenon to which the term "physical reality" can properly be attached'. He is making a pragmatist point here: to attribute a property to an object, for whatever reason, is cognitively meaningful only if the attribution 'makes a difference' so to speak, i.e. only if the attribution implies some definite and discernible practical effects; if it generates *no predictions* about the future, it has no practical consequences, and hence no cognitive meaning. If we measure the position of S_1 and, on the basis of quantum mechanics, attribute a position to S_2 without measuring it, then, Bohr is willing to grant, the position which we thus attribute to S_2 is physically real, since it generates a precise prediction; but he is not prepared to allow that S_2 has in this situation a definite, if unknown, momentum, since the assertion that it has is devoid of practical import.

Recognition of the pragmatist strain in Bohr's philosophy enables us also to comprehend his reply to Frank, which is *prima facie* an avowal of positivism.[78] Here Bohr was merely agreeing with Frank's anti-metaphysical stance towards the notion of the exact simultaneous position and momentum of an electron. Bohr was agreeing that whether or not electrons have such a conjunctive property (even though we cannot measure it) is a metaphysical question in the pejorative sense which positivists are wont to give this term, that is, it is a vacuous question, devoid of factual content since an affirmative answer has no practical consequences. The basis of Bohr's agreement with Frank, however, was not positivism, but pragmatism.

An appraisal of
Bohr's philosophy of physics

Now that my account of Bohr's philosophy of physics has been completed, it remains for me to give a general assessment of it. What value should we assign to Bohr's main ideas – the notions of correspondence and complementarity? How does his interpretation of quantum physics fare in comparison with some of its rivals? What should our assessment of the Bohr–Einstein debate be? I shall take up these questions in the reverse order.

11.1
Einstein or Bohr? The final verdict

The Bohr–Einstein debate is primarily a dispute about the philosophical foundations in terms of which the interpretation of quantum mechanics should be based: specifically it is a dispute about the intrinsic-properties theory, i.e. about the scope of a realist interpretation of quantum mechanics. Einstein wished to maintain the two realist theses, the intrinsic-properties theory and the objective-values theory, whereas Bohr wished to hold only the latter. My own philosophical leanings are in the direction of realism. To my mind the notion of a physical reality the existence and constitution of which are independent of human thought and perception is coherent and plausible: what physical objects there are, and what properties they have, is not determined by human thought or perception, or by human practical interests. I have sympathy, therefore, with Einstein's position concerning the indefinability thesis and the question of the completeness of quantum mechanics.

Einstein argues that, since quantum mechanics retains the mathematical apparatus of the real number system and the classical notions of approximately exact position and approximately exact momentum defined in terms of that system, the notion of approximately exact simultaneous position and momentum is meaningful in quantum mechanics: it is not rendered meaningless by the fact that the conjunctive property which it expresses cannot be measured. Once an appropriate spatio-temporal reference frame has been established, then, with respect to that frame, it makes sense to say that an electron has simultaneously an approximately exact position and an approximately exact momentum, which may be more approximately exact than the uncertainty relation allows; and the fact that this statement has no practical consequences does not render it meaningless. The EPR argument, moreover, indicates that there are physically real properties, such as the property in question, which play no part in a quantum-mechanical description; consequently, there being features of physical reality which quantum mechanics ignores, the theory is in this respect incomplete.

Bohr's counter-argument rests upon a theory of the meaning of physical predicates according to which it is a necessary condition of the meaningful applicability of a physical predicate not only that there be present an instrument suitable for measuring the property in question (the weak meaning condition) but also that a measurement of the property has actually been made (the strong meaning condition). The predicate 'has exact position q and exact momentum p_q at time t' fails to meet these conditions, and hence has no meaningful application. A justification for both of these conditions is required. The strong meaning condition was introduced primarily to save the indefinability thesis from refutation by the EPR argument; Bohr thought that the weak condition alone would not suffice for that purpose. Bohr, however, was mistaken; for in the EPR experiment there is no instrument or collection of instruments capable of being used to measure the exact simultaneous position and momentum of an object. Since the weak condition is not satisfied, Bohr is entitled to hold that even in the EPR experiment, it is not meaningful to talk of the exact simultaneous position and momentum of an object. The strong meaning condition, then, is redundant, and since there are no other grounds for introducing it, save an untenable operationalism or radical empiricism, it ought to be rejected.

A justification of the weak meaning condition is also needed, a justification, that is, for the verificationist element in Bohr's theory of meaning (I am using 'verificationist' as a broad label for theories which equate meaning with assertability conditions). The justification which Bohr would have given, if pressed, would have been based on what I have called the pragmatist strain in his thinking: it is, in a nutshell, that a concept which does no predictive work is devoid of factual content, and hence lacks descriptive meaning. It might be thought that Dummett's verificationist theory of meaning makes available a more sophisticated justification.

According to Dummett, the meaning of a sentence is to be equated not with its truth conditions, but with its assertibility conditions. This view entails that if a sentence lacks assertibility conditions, it lacks a sense. A theory of meaning which is based on a realist conception of truth conditions, Dummett argues, faces severe difficulties. Where the truth conditions are such that, when they obtain, we are able to recognise that they do, there is no problem; but when we are unable to recognise that they obtain, it is difficult to see how we can form any meaningful conception of them. If we cannot recognise the occurrence of the truth conditions, how can we form any comprehensible conception of them? The only answer that Dummett thinks possible is that we form this conception by conceiving of an analogy: we form an imaginative conception of creatures whose cognitive capacities and powers of observation are far superior to ours. For example, we form the conception of the truth conditions of sentences which quantify over an infinite domain by understanding similar sentences which quantify over a finite domain and then by conceiving of some cognitive being who is capable of surveying an infinite domain in a way that is analogous to that in which we may survey a finite domain, and who is able to have direct awareness of the occurrence of the truth conditions in question. Thus we come to assign a meaning to such a sentence by imagining what it would be like to recognise its truth conditions, by imagining the sort of use, albeit unavailable to us, to which such a creature could put the sentence. On this view, in the case of the sentence 'Object O has exact position q and exact momentum p_q at time t', we can comprehend its truth conditions, which admittedly we cannot recognise, by first grasping the truth conditions of the sentences that can be formed from its component predicates, truth conditions which we are able to recognise, and then by imaginatively conceiving of a situation in which both sets of conditions are satisfied and by imagining what it would be like to recognise them.[1]

Dummett does not so much deny that we can in this way come to understand truth conditions the occurrence of which we are unable to recognise (evidence-transending truth conditions, as I shall call them); what he denies is that this understanding can constitute a knowledge of the meaning of the sentence. He argues that (1) to know the meaning of a sentence is to know how to use it correctly for the purposes of communication; hence, (2) if someone knows the meaning of a sentence, he must be able to manifest his knowledge in the appropriate communicative uses of it; but, (3) knowledge of the evidence-transcending truth conditions of a sentence is not a knowledge that can be manifested in the communicative use of the sentence; (4) such knowledge is strictly incommunicable; hence, (5) such knowledge does not constitute knowledge of meaning as an objective and publicly accessible feature of our common language.[2]

Much in this argument is unclear to me. Apropos premiss (2), Dummett does not make clear what he means by 'manifesting' a knowledge of meaning. I take him to mean that to manifest a knowledge of meaning is to show others that one has the knowledge, to act in such a way that others may readily know that one has it. It is not clear what Dummett's grounds are for holding that knowledge of meaning entails the ability to manifest this knowledge. His grounds cannot legitimately be that someone can meaningfully be said to know the meaning of a sentence only if others can know that he knows it, for this would be to assume, question-beggingly, a verificationist principle. Nor can his grounds legitimately be that one has the practical ability to make communicative use of a sentence, and hence knows its meaning, only if one can manifest this ability; for again, this would be tacitly to appeal to a verificationist point, viz. that one can meaningfully be said to have a certain ability only if others can know that one has it. What non-question-begging grounds could there be? The only ones that come to mind are that unless one is able to manifest one's knowledge of meaning, then the meaning in question is incommunicable. But what grounds are there for thinking that one can communicate one's meaning only if one can manifest one's knowledge of it? In order to communicate one's meaning, the hearer must know what one means, but must he also know that one knows what one means? I shall not pursue the question.

But even if it can be shown that knowledge of meaning entails the ability to manifest the knowledge, does it entail the ability to manifest the knowledge in communicative use? By the phrase 'communicative use' Dummett seems to have in mind uses which do not include the giving of a verbal definition. Perhaps, as Dummett argues, it is impossible to give a verbal definition of the meaning of every sentence in the language; but, concerning those sentences the meaning of which *can* be given a verbal definition, must knowledge of their meaning be manifested in communicative uses? Does the giving of a verbal definition not suffice as manifestation of knowledge of the meaning of such sentences? Surely it does. Dummett is perhaps assuming that the meaning of sentences whose meaning is evidence-transcending cannot be given a verbal definition; if so, this ought not to be assumed – an argument for it is required.

What Dummett's grounds are for premiss (3) I do not know. Certainly, knowledge of the evidence-transcending truth conditions of a sentence cannot be manifested in communicative uses of the sentence which are based on recognition of its truth conditions; but it does not follow from this that such knowledge 'cannot, in fact, be fully manifested by actual linguistic practice'.[3] Dummett's grounds, I suggest, are these: if *all* there is to go on is knowledge of evidence-transcending truth conditions, then others cannot tell from the speaker's use of the sentence that he has the requisite knowledge, because, there being no knowledge of assertibility conditions, there is nothing in the speaker's use of the sentence that indicates that he knows what it means, nothing that can serve as a criterion to tell whether he is using the sentence correctly or not. These grounds are persuasive justification of premiss (3).

Premiss (4), however, does not follow from the previous premisses: the fact that knowledge of evidence-transcending truth conditions alone cannot be manifested in communicative use does not entail that such knowledge is incommunicable. Premiss (4) is an additional premiss. Is it not possible, in some cases at least, to manifest knowledge of evidence-transcending truth conditions by means of a verbal definition? The fact (if it is a fact) that it is impossible to give a verbal definition of the meaning of every sentence does not entail that the meaning of sentences the truth conditions of which are evidence-transcending cannot be defined in verbal terms. As Dummett recognises, it is sentences in the more primitive levels of our language that cannot be defined in verbal terms. But sentences the truth conditions of which are evidence-transcending are likely to belong to the higher levels of our language. Dummett, moreover, acknowledges that the ability to give a verbal definition does constitute the ability to manifest a knowledge of meaning.[4] Perhaps a verbal definition of the meaning of such sentences would involve an explication of the sort of analogical conception mentioned above; but nothing in Dummett's argument shows that such an explication would not be an explication of communicable meaning. Dummett may think that the meaning of such sentences cannot be given a verbal definition, perhaps on the grounds that a verbal definition would render the truth conditions of such sentences no longer evidence-transcending.[5] But what could his reasons for this view be? Perhaps that, given a non-trivial definition of the sentence in question (the definiendum), the truth conditions of the sentence which functions as definiens must themselves be evidence-transcending, otherwise the definiendum would not have evidence-transcending truth conditions after all; but then the meaning of the definiens would require a non-trivial verbal definition, and the same consequences as those in the previous case would ensue, and an infinite regress would be generated. But what is vicious about such a regress? Again, I shall not pursue the point.

Dummett's argument for a verificationist theory of meaning, as he presents it, is, I conclude, not sufficiently complete to be regarded as cogent; and hence it does not provide a convincing justification of the verificationist element in Bohr's theory of meaning. Perhaps it can be made cogent; but the onus is on the non-realist to do this.

The verificationist element in Bohr's theory of meaning, then, relies for support on the pragmatist strain in his thinking, in particular on the thesis that if a statement can do no predictive work, it is devoid of factual content. The realist's objection against this thesis is that while a state of affairs can be said to be physically real only if some future consequences are entailed by it, physical reality is not determined by what *we* are able to predict; the pragmatist thesis is too anthropocentric to be tenable. On this point I agree with the realist, and hence with Einstein. If the Bohr–Einstein debate is judged as a philosophical dispute concerning the truth of the intrinsic-properties theory, then Einstein's position is in my view stronger than Bohr's. From a purely philosophical point of view, then, Einstein wins the argument over the question of the completeness of quantum mechanics.

My verdict, however, is not without qualification, for I have some sympathy with Bohr's position. If the property of approximately exact simultaneous position and momentum is one which is ineluctably beyond our powers to measure, then there is little point in complaining about the incompleteness of the theory, since it is pointless to regret ignorance which we are unable to dispel. Some might go further and hold that we cannot properly be said to be ignorant of what we cannot know. I would not go so far as this, since I know of no reason for holding this which is not verificationist. Einstein of course made an issue of the putative incompleteness of the theory simply because he was not convinced that the property in question is ineluctably unmeasurable. Since I agree with Bohr on the measurability question, I see little point in pressing the charge of incompleteness. This is not to admit that the charge is utterly pointless: the point behind it is that quantum mechanics does not rule out the intrinsic-properties theory, as a metaphysical option at least.

Moreover, I have sympathy also with Bohr's reluctance to regard the

notions of exact position and exact momentum, defined in terms of the real
number system, as faithfully mirroring properties of reality as it is in itself.
His conception of these notions is weakly realist in that he regards the
properties which they express as being truly or falsely predicable of objects,
but it is not strongly realist, since he doubts whether these notions provide
faithful pictures of absolute properties which characterise objects as they are
in themselves. Einstein, by contrast, is much less diffident about assigning a
realistic significance to the mathematical structures of physics. However,
Bohr's justifiable reluctance to construe the notions of exact position and
exact momentum in strongly realist terms does not in itself justify his view
that these notions cannot meaningfully be applied to an object simultaneously. Bearing in mind these qualifications, Einstein, I believe, has the
stronger argument, so far as the question of completeness is concerned.

If we take Bell's work into account, however, further qualifications are
needed, since an explanation of Bell's results is forthcoming more readily
from Bohr's interpretation of quantum mechanics than from Einstein's.
Those, like Bohr himself, whose philosophical leanings are towards
pragmatism will be inclined to regard the greater interpretative success of
Bohr's views as clinching the verdict in Bohr's favour; they will argue that
what metaphysical assumptions we adopt should be determined, not by *a
priori* preconceptions, but by the exigencies of the interpretation of our best
scientific theories. If the best interpretation of a highly confirmed theory
requires the rejection of the intrinsic-properties principle, then we should
willingly abandon that principle. Since my own philosophical leanings are
not towards pragmatism, I do not accept this argument. Of course we ought
not to make a dogma of our metaphysics; but we ought not to abandon an
entrenched metaphysical assumption simply because the interpretation of
our latest, and best confirmed scientific theory seems to require it; for
scientific theories, as history shows, come and go, and besides, our
metaphysical assumptions have themselves been adopted in response to the
demands of interpretation of even better confirmed theories than those of
physics, viz. our common-sense view of the world. This is not to say that
physics cannot correct our common-sense views; of course it can and should;
it is to say merely that we should not be too quick to let physics overhaul our
metaphysics. Concerning the case at issue, I am not convinced that the exigencies of interpretation of quantum mechanics call decisively for the
rejection of the intrinsic-properties theory. So far as the question of
the completeness of quantum mechanics – the question of the truth of the

intrinsic-properties theory – is concerned, I think that Einstein was right and Bohr was wrong. However, concerning the question whether some future theory might give a more complete description than quantum mechanics, in that it specifies states hidden to quantum mechanics, in which every observable is assigned an exact value at all times, I think that Bohr was right and Einstein wrong; for I am inclined to think that Bohr was right in holding that kinematic-dynamic complementarity will be a feature of every future theory. Besides, Einstein failed to see that, if it is to be empirically adequate, any future theory which includes hidden states of the intrinsic-properties sort will most likely violate a locality condition: the hidden states will have to be non-locally contextual. There is, then, no outright victor in the Bohr-Einstein contest.

11.2
The notions of correspondence and complementarity

Bohr's notion of correspondence is an important contribution to the philosophy of physics, for it provides the basis for a general account of theory change as a rational process. The notion helps to elucidate the intuitive judgement that a new theory in physics ought to explain both the phenomena explained, and the phenomena unexplained, by the old theory. This condition is generally satisfied through the relation of Bohrian correspondence between the old theory and the new one. What generates the new theory is firstly the recognition and rejection of one of the presuppositions of the old theory, and secondly its replacement by a new hypothesis and a new parameter which, when given a limiting value, enables the laws of the old theory to be derived as limiting cases of the laws of the new theory. The notion of correspondence enables us to do justice both to the continuity and to the discontinuity which are involved in theory change.

Bohr was inclined to present his notion of wave-particle complementarity in a realistic guise. Why? Largely owing to an attraction to a Kierkegaardian view of reality, the multi-aspect view, according to which reality can be fully comprehended only from different points of view or in terms of disparate conceptual schemes. Bohr was well acquainted with the double-aspect theory of the relation between mind and body, and was inclined to view the relation between the wave model and particle model in an analogous light. In one version of the double-aspect theory, mental states and brain states are taken to·be different appearances of one and the same state of the person, different ways in which that state presents itself; in another version (sometimes called the 'two-properties theory'), mental properties and neural properties are taken to be different properties of a person. It is worth noting that Høffding held a version of the double-aspect theory.[6] It is not surprising that Bohr found Spinoza's philosophy congenial, for Spinoza was perhaps the first to propose the double-aspect theory.[7] The classic twentieth-century statement of the two-properties theory is by Samuel Alexander in his *Space, Time and Deity*.[8] It is worth remarking that Bohr got to know Alexander during his time at Manchester; indeed, he took part in the monthly discussion group centred around Alexander and Rutherford.[9]

According to Bohr's more reflective view, however, the non-standard classical models are fictional expedients, of no realistic significance, mere figurative aids to the perceptual imagination of processes that cannot properly be imagined; use of them is optional not mandatory. After he formulated the theory of complementarity in 1927, he continued to speak of the need for a 'renunciation of pictorial representation' as he had been doing since 1920. Properly understood, the notion of wave-particle complementarity is a graphic expression of the fact that electrons cannot be conceived of as classical particles, i.e. as objects which have a definite trajectory. But I cannot see that this fact (if it is a fact) should be given any graphic expression at all, and such a misleading one at that. We ought to resist the desire to visualise what is unvisualisable, not give in to it. Besides, I cannot see that the notion of wave-particle complementarity construed in this way, solves the problem of wave-particle duality. If the trajectory of an electron is thought to be so indefinite as to enable the electron to traverse two slits in a diaphragm simultaneously, then it is difficult to comprehend how the electron can be a compact entity occupying a small spatial volume. Vague talk of an electron's having no definite trajectory does not solve the problem of duality; it merely exchanges one mystery for another. As a solution to the duality problem, the notion of wave-particle complementarity fails.

The notion of kinematic-dynamic complementarity, however, is a different matter entirely. The theory on which the notion is based has two distinct parts, a purely epistemological doctrine including a theory of measurement, and an ontological or semantic doctrine concerning the meaning of physical predicates. Bohr's semantic doctrine is based on an untenable pragmatist theory of meaning. His epistemological doctrine, however, is in my view correct and highly important: it is the only theory I know of which provides an adequate theoretical foundation for the uncertainty principle, and where the notion of wholeness is concerned, the theory has not yet been fathomed in all its depths.

11.3
Alternatives to Bohr's theory of matter and radiation

There are several alternatives to the theory of wave-particle complementarity. They fall into three main classes; instrumentalism, monistic theories, and dualistic theories. According to strong instrumentalism, the notions of the electron, photon etc., do not denote putatively real entities, but are merely devices for generating adequate predictions about the behaviour of certain macroscopic objects. Strong instrumentalism dissolves the problem of duality by eschewing the realist construal of the concepts without which the problem cannot be posed. Weak instrumentalism performs much the same task: on this view the models of particle and wave are held to be very likely false, but it is considered expedient to employ them, so long as they enable us to predict all known phenomena. In view of the intractable nature of the duality problem, instrumentalism in quantum physics must be treated very seriously. It is sometimes said that quantum electrodynamics solves the duality problem, for it shows that a radiation field and a collection of particles can be described either in terms of a wave conception (the quantisation of a classical wave field) or a particle conception (treating the field as a collection of particles and employing the method of second quantisation). This view, however, is mistaken: quantum electrodynamics simply allows wave-particle duality to be given a more systematic theoretical expression than it had in the quantum mechanics of 1926, and it solves the problem of duality only if it is interpreted in instrumentalist terms. The solution of the duality problem, in my view, should be sought not in instrumentalism, but in further developments of quantum mechanics, and especially in an understanding of what quantum-mechanical wholeness really amounts to.

According to monistic theories, electrons and photons are conceived of exclusively either as particle-like entities or wave-like entities as the case may be. In 1926 Schrödinger proposed a monistic interpretation of wave mechanics according to which both radiation and matter consist solely of wave-like entities, discrete energy levels being merely the eigen-frequences of physical waves. He believed that every type of quantum phenomenon could be explained satisfactorily in terms of his wave interpretation. His interpretation faced serious difficulties from the outset, such as the spreading of the ψ-wave of a single electron, and the fact that for a system of n objects the wave-function belongs to a $3n$-dimensional space.[10]

Another sort of monistic theory has been proposed by Alfred Landé, who conceives of matter as being exclusively corpuscular and of radiation as being exclusively undulatory. His thesis is that the behaviour of electrons in the double-slit experiment can be explained satisfactorily by adapting William Duane's corpuscular theory of the diffraction of radiation by an infinite line grating. Duane showed that, given the momentum condition, $\int_0^a p\,dx = nh$, a line grating can take up momentum in a direction orthogonal to the grating only in amounts h/a ('a' signifying the grating constant). He derived a formula for radiation reflexion $(h/c)\,(\sin\theta - \sin\theta') = nh/a$ (with 'θ' signifying the angle of incidence and 'θ'' the angle of reflection) which corresponds to the Bragg formula for X-ray diffraction, $a(\sin\theta - \sin\theta') = n\lambda$.[11] There are, however, serious problems with Landé's attempt to explain electron 'interference'. Duane's theory was generalised to account for Fraunhofer diffraction, but attempts to explain Fresnel diffraction failed, owing to the necessity of attributing to photons the undulatory characteristics of coherence and phase.[12] There is a significant difference between a two-slit system and an infinite periodic lattice structure: in the former case the direction of the particles is orthogonal to, and in the latter parallel to, the direction of periodicity. Besides, the argument, if successful, would support the corpuscular theory of radiation – a consequence which Landé would not have welcomed. Monistic theories have received little support from physicists.

Dualistic interpretations, i.e. conceptions of microphysical objects as in some sense having both corpuscular and undulatory characteristics, are of two main sorts: a diachronic form, according to which corpuscular characteristics and undulatory characteristics are manifested in different situations, at different times, and a synchronic form, according to which microphysical objects have at all times both sorts of characteristics. The diachronic version of the duality thesis is simply the ontic interpretation of wave-particle complementarity. It has little to recommend it: the hypothesis that radiation is undulatory when freely propagating, and corpuscular when interacting with matter, creates the further problem of explaining the sudden transformations of characteristics – a problem which a realist must take seriously. Moreover, whereas some matter–matter interactions call for the wave model (e.g. the Davisson–Germer experiment), others call for the particle model (e.g. Wilson cloud-chamber experiments). Interpretation in terms of both the particle model and the wave model is thus too piecemeal and adventitious to be satisfactory, there being no systematic theory of the circumstances in which each model is applicable. For a realist, the *ad hoc* character of a dualist interpretation must be regarded as a serious drawback.

De Broglie's 'double-solution' theory of 1927 is a dualistic theory of the synchronic type, in which microphysical objects are conceived of as corpuscle-like concentrations of energy within a physical wave which determines their motion. Following an unsympathetic reception, de Broglie abandoned his theory, though he took it up again in the early fifties after the publication of Bohm's dualistic theory, according to which the wave function describing a particle is taken to represent a real physical field which exerts a force on the particle, thus determining its motion.[13] De Broglie's and Bohm's theories provide an explanation of 'interference' effects: in the two-slit experiment the particle goes through one slit, and its associated 'ghost' wave through both slits, the two parts recombining and interfering with each other and guiding the particle away from certain areas of the screen. These theories, however, are not so much interpretations of quantum mechanics as substantive modifications of it, for they contain postulates additional to those of orthodox quantum mechanics. Few physicists have considered them worth pursuing, no doubt because they seem irremediably *ad hoc*: Bohm's theory for example is *ad hoc* in that it is not independently testable, yielding no predictions which differ from those of quantum mechanics, and its additional postulates are introduced solely to secure agreement with the predictions of quantum mechanics. The ψ-field for example is required to have some curious characteristics: since it is neither radiated nor absorbed, it is undetectable, which is why it is called a 'ghost field'; and it takes on infinite values at certain points – if the infinite values are positive, the field repels the particle from the point; if negative, the particle passes the point with infinite velocity. The latter condition is required in order to account for the two-slit experiment.[14] A testable theory of the synchronic kind has been proposed by R.D. Prosser, who suggests a model according to which radiation is at once a collection of infinite waves determining the undulatory characteristics and a wave-packet determining the corpuscular characteristics. In terms of this model Prosser is able to explain not only 'interference' phenomena, without having to resort to the notion of mutual destruction and re-inforcement of superimposed waves, but also corpuscular phenomena such as the photo-electric effect and the Compton effect.[15] Prosser's theory has not attracted a great deal of interest; nevertheless it is the sort of theory that might supplant orthodox quantum mechanics, providing a solution to the intractable problem of wave-particle duality.

Since none of the above theories is entirely satisfactory, it may be that we shall have to postulate some sort of action at a distance in order to explain the double-slit experiment. If in this experiment a detector is placed behind one of the slits, it is possible to tell, before it hits the screen, through which slit a particle has passed: if the detector were not triggered, then the particle must have passed through the other slit. If we place the detector alternately behind one slit and then behind the other, and gather together all the instances in which a particle passed through the slit that has no detector, then these instances will not produce the interference pattern on the screen. Thus, the behaviour of the particles that pass through one slit differs depending on whether or not a detector is placed at the slit through which they do not pass. This curious phenomenon suggests that the presence of a detector, or a slit, at a place where particles do not pass has a causal influence on their behaviour; it suggests some sort of action at a distance, though not necessarily an action which spans empty space. Bohr's notion of wave-particle complementarity cannot adequately explain this curious phenomenon; nor can his idea of wholeness as it stands. Since, however, an organic whole can be conceived of as a system of causal inter-relations, the idea of wholeness may be a fruitful basis on which to build a theory of an appropriate action at a distance.

11.4
Many worlds and quantum logic

Among realist interpretations of quantum mechanics other than Einstein's, two in particular stand out: the many-worlds interpretation and the quantum logic interpretation. According to Henry Everett's interpretation, the state vector describes an individual object, and a superposition of state vectors describes a collection of individual objects which belong to different real worlds. The superposed vector describing the system of the object and the instrument is transformed by the process of measurement into a new superposed vector, the component vectors of which refer to different real worlds. Measurement has the effect of generating new real worlds which do not communicate with each other. Leaving aside matters of detail, the main disadvantage of this interpretation is its ontological extravagance: it multiplies entities beyond necessity – though this extravagance may not embarrass someone who holds a realist interpretation of possible-worlds semantics for modal logic. Bohr's theory of kinematic-dynamic complementarity, however, is preferable to the many-worlds interpretation, since it secures the same benefits without the expense of a superabundant ontology. Philosophically interesting though it is, the many-worlds interpretation has not attracted much support.[16]

The quantum logic interpretation has attracted much wider interest. The thesis is that the experimental statements of quantum mechanics, construed as describing the physical properties of objects, are governed by a non-classical propositional calculus which has a non-Boolean algebraic structure similar to the non-Boolean structure of the set of subspaces of a Hilbert space (I shall confine my discussion to versions of quantum logic which are based on the set of subspaces). The basis of the interpretation is a certain correspondence between the experimental statements and the subspaces: there is a many-to-one mapping of the statements into the set of subspaces, and a one-to-one mapping of equivalence classes of statements into the set of subspaces. The set of subspaces may be described in terms of different algebraic structures. The strongest of these, to which I shall confine my remarks, is a lattice, i.e. a partially ordered set in which each pair of elements has a greatest lower bound ('GLB') and a least upper bound ('LUB'). There is also a correspondence between compound experimental statements and subspaces. The conjunction $a \wedge b$ of the statements a, b corresponds to the GLB of the subspaces corresponding to a and b in the lattice of subspaces; the disjunction $a \vee b$ corresponds to the LUB of the corresponding subspaces; and the negation $-a$ corresponds to the orthogonal complement of the subspace corresponding to a. Quantum logic is the calculus which results from this mapping. Since the lattice of subspaces is non-Boolean, so too is quantum logic. The lattice of subspaces is non-Boolean since the formulas for the distribution of conjunction over disjunction, and disjunction over conjunction,

$$a \wedge (b \vee c) = (a \wedge b) \vee (a \wedge c),$$
$$a \vee (b \wedge c) = (a \vee b) \wedge (a \vee c),$$

are not theorems or semantically valid in the lattice of subspaces. How does this come about?

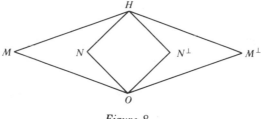

Figure 8

A countably-infinite-dimensional Hilbert space forms a σ-complete, atomic, non-distributive, non-modular, orthomodular lattice. It is the presence of incompatible subspaces which leads to the failure of distributivity. Two subspaces M, N are incompatible if and only if, together with their orthogonal complements $M^{\perp}$, $N^{\perp}$, they do not form a distributive sublattice. Two subspaces M, N are compatible if and only if there are three mutually orthogonal subspaces M_I, N_I, L such that $M = M_I \vee L$ and $N = N_I \vee L$. The lattice of incompatible subspaces M, N can be represented as shown in Figure 8. It is the presence of subspaces which are not mutually orthogonal that gives rise to non-distributivity in the lattice of subspaces. For example, if M, N and $N^{\perp}$ are subspaces of a two-dimensional Hilbert space which correspond respectively to the experimental statements $a = $ 'The x-axis spin of electron E in state ψ at time t is $+\frac{1}{2}\hbar$,' $b = $ 'The z-axis spin of electron E in state ψ at time t is $+\frac{1}{2}\hbar$,' and $c = $ 'The z-axis spin of electron E in state ψ at time t is $-\frac{1}{2}\hbar$,' then the lattice consisting of M, N and their orthogonal complements is non-distributive, since $M \wedge (N \vee N^{\perp}) = M \wedge H = M$, whereas $(M \wedge N) \vee (M \wedge N^{\perp}) = 0 \wedge 0 = 0$, and $M \neq 0$ (H and 0 are the unit and zero elements of the lattice). The lattice is non-distributive since M is not orthogonal to N or $N^{\perp}$: M and N are incompatible subspaces.

In his earlier treatment of the quantum logic interpretation, Putnam argued that quantum logic and classical logic are mutually incompatible rivals: just as it is an empirical question what pure geometry correctly describes the mathematical structure of physical space, so also it is an empirical question which propositional calculus correctly describes the logical structure of our discourse; and just as non-Euclidean geometry has supplanted Euclidean geometry in the description of physical space, so also may quantum logic supplant classical logic in the formalisation of our discourse.[17] On Putnam's thesis it is possible that some well-formed, syntactically valid formula of classical logic, such as one of the distributive equivalences, should turn out to be false under some substitution of statements for its statement variables; and this possibility, he suggests, is in fact realised for some substitutions of experimental statements of quantum mechanics.

Putnam's thesis entails that quantum logic has a different semantics from that of classical logic, for, given the classical two-valued semantics, it is not conceivable that the distributive equivalences should turn out to be false on some substitution of statements for the statement variables; given these semantics, the distributive equivalences necessarily are semantically valid. The classical semantics are not suitable for quantum logic, since the set of tautologies which they generate does not match the set of theorems. Putnam's thesis, moreover, presupposes that the meanings of the classical logical connectives are preserved under the mapping into the lattice of subspaces, that is, that the connectives of quantum logic can be assigned the same natural-language meanings as those of classical logic; otherwise it would make no sense to say that quantum logic contradicts classical logic. This being the case, Putnam is constrained to deny that the meanings of logical connectives are determined by their formal semantics, for, if the meanings are determined by the semantics, then the meanings of the classical logical connectives are different from those of the quantum logical connectives, the respective semantics being different. This denial is, to say the least, extremely implausible. But even if Putnam maintains that the meanings of logical connectives are determined, not by their formal semantics, but by the set of axioms or rules of inference of the logics to which they belong, his view faces the further difficulty that classical logic and quantum logic do not share the same set of axioms or rules of inference. Putnam's solution to this difficulty is to say that while the connectives of the two logics are not entirely synonymous, they have the same 'core meanings', and hence are not so different in meaning as to preclude the mutual incompatibility of the two logics. The disjunction, conjunction, and negation connectives in the two logics have the same core meanings determined by the following set of theorems which the two logics have in common:

(1) a implies $a \lor b$.
(2) b implies $a \lor b$.
(3) if a implies c, and b implies c, then $a \lor b$ implies c.
(4) a, b together imply $a \land b$.

(5) $a \wedge b$ implies a.

(6) $a \wedge b$ implies b.

(7) $-(a \wedge -a)$ holds.

(8) $(a \vee -a)$ holds.

(9) $-(-a)$ is equivalent to a.[18]

I take it that the core meaning of a connective is the basic meaning which is essential to its being regarded as a disjunction connective, or conjunction connective, etc. Apropos disjunction, however, one might say more plausibly that the core meaning is given by the formulas,

$$a \vee b \text{ implies } a \text{ or } b,$$
$$(a \vee b) \wedge -a \text{ implies } -b,$$

than by the formulas (1)–(3) in Putnam's list. But these two formulas are not theorems of quantum logic. *Pace* Putnam, the failure of the distributive equivalences is *not* the only difference between classical logic and quantum logic.

Since the formula '$a \vee b$ implies a or b' is not an axiom or theorem of quantum logic, '$a \vee b$' cannot be interpreted as meaning 'at least one of a, b is true': '$a \vee b$' may be true, though neither a nor b is true. The difference in meaning between the classical disjunction connective and the quantum connective '$\vee$' is much greater than Putnam imagines. The difference can be traced to the important distinction between classical negation and the negation operation in quantum logic. Whereas classical negation is analogous to set-theoretical complementation, the negation operation in quantum logic is analogous to lattice orthocomplementation. The orthocomplement $M^{\perp}$ of a subspace M is not the difference between the whole space H and the space M but the set of subspaces orthogonal to M. The lattice of subspaces of Hilbert space is not closed with respect to set complementation. Classical negation, sometimes called 'exclusion negation', simply removes a proposition from the set of true propositions, whereas negation in quantum logic is what is sometimes called 'choice negation': $a^{\perp}$ implies that one of the set of determinable alternatives to a (determined by the context) is true. This difference is important, since 'disjunction' in quantum logic is defined in terms of negation and conjunction by way of the de Morgan equivalence: $a \vee b = \mathrm{def} -(-a \wedge -b)$. The 'disjunction' functor in quantum logic is similar, not to set-theoretical union, but to lattice-theoretical LUB. The LUB of two subspaces M, N is the span of M and N, i.e. the topological closure of the sum of vectors in M and N, which, unlike the union, may contain vectors which are neither in M nor N; span is equivalent to union only if M and N are compatible subspaces. Owing to this difference, the classical rule '$-(a \wedge b) \wedge a$ implies $-b$' does not hold in quantum logic; for although two subspaces M, N are disjoint in that $M \wedge N = 0$, it is not necessarily the case that $M \leqslant N^{\perp}$ or $N \leqslant M^{\perp}$ – the latter holds only if M and N are compatible subspaces.

There are, then, no good grounds for thinking that the meanings of the connectives of classical logic and quantum logic are sufficiently similar to make it plausible to say that quantum logic is incompatible with classical logic. Quantum logic is an alternative to classical logic only in the inclusive sense, and not in the exclusive sense. Consequently, by adopting quantum logic one resolves the paradoxes of quantum mechanics, if at all, merely by changing the subject.

One may of course regard quantum logic as an inclusive alternative to classical logic, and one which provides a more adequate formal representation of the logical relations between quantum-mechanical predicates. This is the view that is taken by members of the University of Western Ontario school. Bub, for example, maintains that quantum mechanics should be interpreted realistically, as a theory that describes real physical events: it specifies certain 'structural constraints' on properties and events; it describes how the physical properties of an object 'hang together', 'what possible events are open to it'. Moreover, the quantum-mechanical properties and events 'hang together' in a non-classical, non-Boolean way.[19] I take Bub's obscure remarks to mean that quantum mechanics describes the logical relations between the physical properties of an object. If this reading is correct, then Bub's thesis is that the physical predicates of quantum mechanics are governed not by the classical predicate calculus, but by the quantum-logical predicate calculus, and consequently that the experimental statements of quantum mechanics are governed by quantum logic.

There are at least two major problems with a view such as Bub's. The first concerns the relation of incompatibility between predicates, the relation which corresponds to the relation of incompatibility between subspaces. How is this relation to be understood? If one gives quantum logic a realist semantics, as distinct from a verificationist or operationalist semantics – i.e. a semantics in which truth is construed in realist terms – then the relation is treated as a relation of logical incompatibility between physical predicates – I shall call this the 'incompatibility thesis'. Now, those who adopt a realist construal of quantum logic generally wish also to hold the objective-values thesis to the effect that measurement reveals the pre-existing value of an observable (to say nothing of the intrinsic-values thesis). But there is a problem in holding both the incompatibility thesis and the objective-values thesis. If we measure, say, the position of an object O at time t, and find the value x, then x was the position of O immediately before t. If position and momentum, however, are logically incompatible properties, then O did not possess a definite momentum at t. The difficulty is to reconcile the objective-values thesis with the ostensible fact that before t we were free to measure either the position or the momentum of O, but not both – the 'free-choice thesis' as I shall call it. The free-choice thesis and the incompatibility thesis together exclude the objective-values thesis, and call instead for the creation-by-measurement thesis. But the creation-by-measurement thesis is incompatible with the main motive for adopting the realist construal of quantum logic, which is to uphold a realist construal of the experimental statements, according to which these statements have to do not just with measured values of observables but also with pre-existing values. If one maintains the objective-values thesis (a fairly weak realist assumption), then one must abandon either the free-choice thesis (which seems impossible) or the incompatibility thesis (which seems equally impossible if one is to maintain the realist construal of quantum logic.) Bub seems unaware of the difficulty, which I shall call the 'coherence problem'.

In his later writings on the topic, Putnam is aware of the problem. The realist may respond, he observes, by maintaining that the coherence problem arises only if we construe the incompatibility relation in classical terms, which of course we ought not to do, since the relation is non-Boolean. The non-Boolean character of the relation seems to consist in the possibility that two propositions a, b should each be true yet their conjunction should not be true. Putnam denies that this conception of incompatiblity is intelligible, and so he rejects a realist construal of quantum logic, one that involves a semantics in which the notion of truth is understood as a relation of correspondence between a statement and independently existing reality. Instead he opts for a non-realist semantics, in which truth is defined in terms of verification conditions. On this view a statement in quantum logic is true (or correct) if it is verified by performing a measurement in which the state vector of the object lies in the subspace of Hilbert space corresponding to the value specified by the statement.[20] Thus two statements a, b may be incompatible in quantum logic, since their conjunction $a \wedge b$ is not verifiable, while each individually is true. In his later work Putnam recognises that conjunction in quantum logic does not have the same sense as that in classical logic.

I agree with Putnam that no sense can be made of the notions of incompatibility and conjunction in quantum logic if these are construed in realist terms. But his non-realist construal of quantum logic, while disposing of the coherence problem, has no advantage over Bohr's interpretation of quantum mechanics. Putnam recognises that his quantum logic interpretation resembles one version of the Copenhagen interpretation (it is in fact Bohr's version). What Putnam does is to build a new logic around the incompatibility relation. There is, however, no need to do this, since the relation can be understood perfectly well in classical terms if one explicates it in terms of Bohr's theory of kinematic-dynamic complementarity. Nor are there any explanatory advantages in adopting Putnam's approach: it does not make quantum mechanics more comprehensible; indeed, it provides no explanation of the relation of incompatibility, and merely builds the puzzling features of quantum mechanics into the logic in terms of which the theory is formalised.

The second major problem is to assign comprehensible natural-language meanings to the connectives of quantum logic, a problem that is particularly acute in the case of disjunction. In what sense of 'or' can '$a \lor b$' be interpreted to mean 'Either a or b', if '$a \lor b$' may be true when neither a nor b is true? Bub does not address this problem. Putnam, again, is aware of the problem, and thinks that it is alleviated if quantum logic is given a verificationist semantics. A disjunction in quantum logic can be true, since we can specify a verification for it as a whole, while the disjuncts individually are not true, there being no verification conditions for these taken individually.(ib.) But he does not say what the relevant measurement operations are; and so it is far from clear that the problem is in fact alleviated. I shall not pursue this point, however, since my concern is to expose the weaknesses in the realist construal of quantum logic rather than to criticise non-realist construals.

Allen Stairs is aware of both problems, and believes that both problems can be solved. Apropos the second problem, Stairs bites the bullet and holds that a quantum-logical disjunction may be true even though none of the disjuncts is true. To make sense of this, he suggests, we must make sense of the idea of a disjunctive fact, a fact which does not depend on the truth of the component disjuncts. An example of such a fact is the state of affairs in which a pebble seems to belong to one or other of two neighbouring piles of pebbles without clearly belonging to either pile. We may be prepared confidently to assign the value 'true' to the disjunction 'Either pebble x belongs to pile A or pebble x belongs to pile B', while declining to assign either disjunct the value 'true': we may hold that these disjuncts have no definite truth-value, or a third truth-value such as 'neither true nor false'.[21] This is an interesting suggestion, and it remains to be seen whether it can be made good, whether a quantum logic with an appropriate semantics can be developed which can be given a comprehensible interpretation in natural-language terms. I do not believe that an appropriate three-valued semantics can be given which captures the algebraic structure of the lattice of subspaces of a Hilbert space, but I will not pursue this point here.

Apropos the first problem, the coherence problem, Stairs recognises that a strong realist version of the quantum logic interpretation must be abandoned. Stairs' solution to the problem is to abandon the intrinsic-properties theory, according to which all observables of an object have definite values at any time. Stairs proposes instead that the values of some quantum-mechanical observables are indefinite. This proposal of course is not new, but the use to which Stairs puts it is new. By relinquishing the intrinsic-properties theory, Stairs is able to retain the principle of local action. Consider the quantum-mechanical inequality,

$$\text{prob}(A+,C+) \geqslant \text{prob}(A+,B+) + \text{prob}(B+,C+).$$

If we adopt the indefinite-values theory, this inequality should correspond to the following statistical inequality (where the asterisk stands for an indefinite pre-existing value):

$$N(+*-,-*+) \geqslant N(+-*,-+*) + N(*+-,*-+).$$

The latter inequality is unproblematical: it can be construed as an expression of the fact that pairs with pre-existing values of the type $(+*-,-*+)$ may occur much more frequently than pairs with pre-existing values of the types $(+-*,-+*)$ and $(*+-,*-+)$. There is no need to invoke the idea of an action at a distance in order to explain this inequality.

Several objections can be raised against Stairs' proposed solution to the coherence problem. First, his proposal constitutes a serious departure from the original intention behind the quantum logic interpretation, viz. a realist interpretation of quantum mechanics according to which every quantum-mechanical observable has a definite value at all times. Second, the objective-values theory must also be abandoned. For, since measurement always reveals definite values, and never indefinite ones, it would be stretching coincidence too far to suggest that whenever we make a measurement, we happen invariably to measure definite pre-existing values. It might be thought that Bohr faces the same difficulty, but this is not so, since he could explain why it is that measurement always reveals definite values: the explanation is that the value of an observable to be measured becomes definite as soon as the appropriate measuring instrument has been set up. Stairs indeed recognises that he is constrained to give up the objective-values theory, and he explicitly adopts the creation-by-measurement theory, an exceedingly obscure theory. My main objection, however, is that the abandonment of the objective-values theory constitutes a further departure from the intention behind the realist construal of the quantum logic interpretation: contrary to the original intention, measurement plays a specific role in the interpretation of quantum mechanics after all.[22] The abandonment of the intrinsic-properties theory and the objective-values theory is a severe blow to the realist construal: that construal is realist only in a Pickwickian sense.

Third, Stairs recognises that his modified quantum logic interpretation resembles the Copenhagen interpretation, though he plays down the similarity. Whereas on the Copenhagen interpretation unmeasured observables cannot meaningfully be said to have definite values, on the modified quantum logic interpretation unmeasured observables may be said to have indefinite values. This difference, however, is not as great as Stairs maintains.[23] His view amounts simply to the ontic construal of the state vector (according to which a superposition of state vectors describes an ontically indefinite state), together with the creation-by-measurement theory; the only substantially new feature is the thesis that the experimental statements of quantum mechanics are governed by quantum logic, and not by classical logic. What advantage does Stairs' proposed interpretation have over the Copenhagen interpretation thus described? The advantage which Stairs claims is that his interpretation is at least closer to a realist point of view than is the Copenhagen interpretation. Doubtless this is true, but the advantage is bought at a price, an abandonment of classical logic; and it is not clear that the goods can be delivered, since a formal semantics that can be interpreted in natural language terms remains to be worked out. The quantum logic interpretation, then, is more a programme for an interpretation than an interpretation itself. Besides, the advantage which Stairs' proposed interpretation has over the Copenhagen interpretation can be had at less cost if we adopt Bohr's interpretation, for the latter at least secures the objective-values theory, and that without incurring the expense of subscribing to a new, and not fully developed logic. Furthermore, Stairs is required to explain what ontic indefiniteness consists in, and why certain observables in quantum mechanics have ontically indefinite values, and why it is that these values become definite when they are measured. Now Bohr's interpretation can provide answers to these questions from its own resources. If we wish to talk of ontic indefiniteness, and Bohr prefers that we do not, then ontic indefiniteness consists in the absence of the physical conditions which constitute the preconditions for the instantiation of the properties in question, and indefinite values become definite simply because the appropriate preconditions have been satisfied. Bohr's interpretation, then, has greater explanatory power than the quantum logic interpretation; moreover, it delivers a greater degree of realism, and all this without putting us to the expense of a major revision of classical logic. Bohr's interpretation of quantum physics is superior to its realist rivals; it is superior also to its less realist rivals such as instrumentalism; even taking into account its weak points – the thesis of wave-particle complementarity and the indefinability thesis – it is an interpretation of considerable explanatory power and philosophical subtlety.

NOTES

Published works cited in the Notes are referred to by the author's name, title, publisher, place and year of publication at their first occurrence; thereafter only the name of the author and a short title are given. The following abbreviations are employed:

AHQP	Archive for the History of Quantum Physics
Ann. Phys.	*Annalen der Physik*
APHK	Bohr, *Atomic Physics and Human Knowledge*
Arch. Hist. Ex. Sci.	*Archive for History of Exact Sciences*
ATDN	Bohr, *Atomic Theory and the Description of Nature*
Brit. J. Phil. Sci.	*British Journal of the Philosophy of Science*
Coll. Works	*Niels Bohr: Collected Works*
Compt. Rend. Acad. Sci.	*Comptes Rendus de l'Académie des Sciences*
EPR	Einstein, Podolsky and Rosen
Essays	*Essays, 1958–1962 on Atomic Physics and Human Knowledge*
Hist. Stud. Phys. Sci.	*Historical Studies in the Physical Sciences*
Internat. J Th. Phys.	*International Journal of Theoretical Physics*
J. Chem. Soc.	*Journal of the Chemical Society*
J. Phil.	*Journal of Philosophy*
Naturw.	*Naturwissenschaften*
NBA: BMSS	Niels Bohr Archive: Bohr Manuscripts
NBA: BSC	Niels Bohr Archive: Bohr Scientific Correspondence
Pauli: Wiss. Brief.	*Wolfgang Pauli: Wissenschaftlicher Briefwechsel mit Bohr, Einstein, Heisenberg u.a., Band 1: 1919–1929*
Phil. Mag.	*The Philosophical Magazine*
Philos. Trans. Roy. Soc. Lond.	*Philosophical Transactions of the Royal Society (London)*
Phys. Rev.	*The Physical Review*
Phil. Sci.	*Philosophy of Science*
Phys. Z.	*Physikalische Zeitschrift*
Proc. Arist. Soc.	*Proceedings of the Aristotelian Society*
Proc. Cam. Phil. Soc.	*Proceedings of the Cambridge Philosophical Society*
Proc. Nat. Acad. Sci.	*Proceedings of the National Academy of Sciences*
Proc. Phys. Soc.	*Proceedings of the Physical Society*
Proc. Roy. Soc. Lond.	*Proceedings of the Royal Society (London)*

Revs. Mod. Phys.	*Reviews of Modern Physics*
Sitz. preuss. Akad. Wiss.	*Sitzungsberichte der Königlichen Akademie der Wissenschaften zu Berlin, phys.-math. Klasse*
Sitz. Wiener Akad. Wiss.	*Sitzungsberichte der Wiener Akademie der Wissenschaften, math.-nat. Klasse*
Stud. Hist. Phil. Sci.	*Studies in the History and Philosophy of Science*
Z. Phys.	*Zeitschrift für Physik*
Verh. dt. Phys. Ges.	*Verhandlungen der Deutschen Physikalischen Gesellschaft*

Chapter 1
Wave-particle duality

1 See M. Planck, 'Über eine Verbesserung der Wien'schen Spektralgleichung', *Verh. dt. Phys. Ges.*, **2**, 202–4, 1900; 'Zur Theorie des Gesetzes der Energieverteilung im Normalspektrum', *Verh. dt. Phys. Ges.*, **2**, 237–45, 1900 (both translated in *The Old Quantum Theory*, ed. D. ter. Haar, Pergamon, Oxford, 1967). See also T.S. Kuhn, *Black-Body Theory and the Quantum Discontinuity: 1894–1912*, Clarendon Press, Oxford, 1978; J. Mehra & H. Rechenberg, *The Historical Development of Quantum Theory*, Springer-Verlag, New York, 1982, Vol. I, Ch. 1.

2 M. Planck, 'Zur Theorie des Gesetzes der Energieverteilung', p. 239.

3 See M. Planck, 'Über die Beziehung zwischen dem zweiten Hauptsätze der mechanischen Wärmetheorie und der Wahrscheinlichkeitsrechnung respektive den Sätzen über das Wärmegleichwicht', *Sitz. Wiener Akad. Wiss.*, **76**, 1877, pp. 376f.

4 See T.S. Kuhn, *Black-Body Theory*, pp. 125–30.

5 See Einstein, 'Über einen die Erzeugung und Verwandlung des Lichtes betreffenden heuristischen Gesichtspunkt', *Ann. Phys.*, **17**, 1905, pp. 133, 143.

6 See Lord Rayleigh (R.J. Strutt), 'The Dynamical Theory of Gases and Radiation', *Nature*, **72**, 54–5, 1905.

7 Lord Kelvin (W. Thomson), 'Nineteenth-century Clouds over the Dynamical Theory of Heat and Light', *Phil. Mag.*, **2**, 1–40, 1901; see also Kelvin, 'A decisive Test Case disproving the Maxwell-Boltzmann Doctrine regarding the Distribution of Kinetic Energy', *Phil. Mag.* (*Ser.* 5), **33**, 466–7, 1892. See also S.G. Brush, 'Foundations of Statistical Mechanics, 1845–1915', *Arch. Hist. Ex. Sci.*, **4**, 1967, pp. 162ff.

8 See J. Jeans, 'A Comparison between two Theories of Radiation', *Nature*, **72**, 1905, p. 294.

9 See Einstein, 'Zur Theorie der Lichterzeugung und Lichtabsorption', *Ann. Phys.*, **20**, 199–206, 1906. The formalism in the text above is modernised.

10 See C. Seelig, *Albert Einstein: a Documentary Biography*, trans. M. Savill, Staples Press, London, 1956, p. 74.

11 Einstein, 'Über einen die Erzeugung und Verwandlung des Lichtes', p. 143.

12 Ibid.

13 Einstein, 'Über einen die Erzeugung und Verwandlung des Lichtes', p. 132.

14 See P. Lenard, 'Erzeugung von Kathodstrahlen durch ultraviolettes Licht', *Ann. Phys.*, **2**, 1900, pp. 374f. For other classical explanations of the photo-electric effect see R.H. Stuewer, 'Non-Einsteinian Interpretations of the Photo-electric Effect', *Historical and Philosophical Perspectives of Science*, ed. R.H. Stuewer, University of Minnesota Press, Minneapolis, 1979, pp. 246–63. See also B.R. Wheaton, *The Tiger and the Shark: Empirical Roots of Wave-Particle Dualism*, Cambridge University Press, Cambridge, 1983, pp. 74ff.

15 See Einstein, 'Über einen die Erzeugung und Verwandlung des Lichtes', p. 132.

16 See Einstein, 'Über die Entwicklung unserer Anschauungen über das Wesen und die Konstitution der Strahlung', *Phys. Z.*, **10**, 1909, pp. 819, 821.

17 See Einstein, 'Zur Theorie der Lichterzeugung', esp. pp. 199, 202f.

18 See A. Hermann, *The Genesis of the Quantum Theory, 1899–1913*, trans. C.W. Nash, M.I.T. Press, Cambridge, Mass., 1971, pp. 35ff.; see also T.S. Kuhn, *Black-Body Theory*, pp. 189ff.; E. Garber, 'Some Reactions to Planck's Law, 1900–1914', *Stud. Hist. Phil. Sci.*, **7**, 89–126, 1976.

19 See A. Hermann, *The Genesis of the Quantum Theory*, pp. 41ff.

20 See Einstein, 'Zum gegenwärtigen Stand des Strahlungsproblems', *Phys. Z.*, **10**, 1909, p. 189.

21 See Einstein, 'Zum gegenwärtigen Stand', pp. 192f.

22 See Einstein, 'Zum gegenwärtigen Stand', pp. 817, 826.

23 See Einstein, 'Zum gegenwärtigen Stand', p. 826f.

24 See H.A. Lorentz, 'Die Hypothese der Lichtquanten', *Phys. Z.*, **11**, 349–54, 1910. See Einstein's letter of 23 May 1909 to Lorentz (cited in R.H. Stuewer, *The Compton Effect: the Turning Point in Physics*, Science History, New York, 1975, pp. 48ff.); see also R. McCormmach, 'Einstein, Lorentz, and the Electron Theory', *Hist. Stud. Phys. Sci.*, **2**, 1970, pp. 74f.

25 See Einstein's letters to Besso, *Albert Einstein–Michele Besso Correspondance, 1903–1955*, ed. P. Speziali, Hermann, Paris, 1972, pp. 19, 130.

26 See A.L. Hughes, 'On the Emission Velocities of Photoelectrons', *Philos. Trans. Roy. Soc. Lond.*, **212**, 205–26, 1912; and O.W. Richardson & K.T. Compton, 'The Photoelectric Effect', *Phil. Mag.*, **24**, 575–94, 1912.

27 See R.A. Millikan, 'A direct photoelectric Determination of Planck's *h*', *Phys. Rev.*, **7**, 355–88, 1916; see also R.H. Kargon, 'The conservative Mode: Robert A. Millikan and the twentieth-century Revolution in Physics', *Isis*, **68**, 509–26, 1977.

28 See R.A. Millikan, 'A direct photoelectric Determination', pp. 382, 384. See R.H. Stuewer, *The Compton Effect*, pp. 74f.

29 See Einstein, 'Quantentheorie der Strahlung', *Phys. Z.*, **18**, 1917, p. 121. See also M.J. Klein, 'Einstein and the Wave-particle Duality', *The Natural Philosopher*, **3**, 3–49, 1963.

30 Einstein, 'Quantentheorie der Strahlung', p. 128.

31 Einstein, 'Quantentheorie der Strahlung', p. 121.

32 See Einstein's letter of 6 September 1916 to M. Besso, *Albert Einstein–Michele Besso Correspondance*, p. 82.

33 See Einstein's letter of 30 January 1921 to M. Born, *The Born-Einstein Letters*, ed. M. Born, MacMillan, London, 1971, p. 50.

34 See Einstein, 'Über ein den Elementarprozess der Lichtemission betreffendes Experiment', *Sitz. preuss. Akad. Wiss.*, 1921, pp. 882–3; see also M.J. Klein, 'The First Phase of the Bohr–Einstein Dialogue', *Hist. Stud. Phys. Sci.*, 2, 1970, pp. 8ff.

35 See Einstein's letter to M. Born, *The Born–Einstein Letters*, pp. 65, 71.

36 See Einstein's letter of 19 January 1922 to P. Ehrenfest (cited in M.J. Klein, 'The First Phase', p. 12.) See also Einstein, 'Zur Theorie der Lichtfortpflanzung in dispergierenden Medien', *Sitz. preuss. Akad. Wiss.*, 1922, pp. 18–22.

37 See A.H. Compton, 'The Softening of Secondary X-rays', *Nature*, 108, 366–7, 1921; see R.H. Stuewer, *The Compton Effect*, pp. 172ff.

38 See R.H. Stuewer, *The Compton Effect*, pp. 190ff; see A.H. Compton, 'Secondary Radiations produced by X-rays, and some of their Applications to Physical Problems', *Bulletin of the National Research Council*, 1922, pp. 18f, 56 (cited by R.H. Stuewer, *The Compton Effect*, p. 211).

39 See A.H. Compton, 'A Quantum Theory of the Scattering of X-rays by light Elements', *Phys. Rev.*, 21, 483–502, 1923 (received 13 December 1922, published May 1923).

40 See R.H. Stuewer, *The Compton Effect*, Chs. 3–6.

41 See C.T.R. Wilson, 'Investigations on X-rays and Beta-rays by the Cloud Method. Part 1: X-rays', *Proc. Roy. Soc. Lond.*, 104, 1–24, 1923; and W. Bothe, 'Über eine neue Sekundärstrahlung der Röntgenstrahlen: 1 Mitteilung', *Z. Phys.*, 16, 319–20, 1923. See R.H. Stuewer, *The Compton Effect*, pp. 242f.

42 See A.H. Compton, Note to *Nature*, 112, 1923, p. 435.

43 See R.H. Stuewer, *The Compton Effect*, pp. 246ff, 290f.

Chapter 2

Niels Bohr and wave-particle duality

1 See Bohr, *Studier over Metallernes Elektrontheori*, 1911, translated in *Niels Bohr: Collected Works*, gen. ed. L. Rosenfeld, Vol. I, Early Work (1905–11), ed. J. Rud Nielsen, North-Holland, Amsterdam, 1972, pp. 297–395.

2 See Bohr, 'On the Constitution of Atoms and Molecules', *Phil. Mag.*, 26 1–25, 476–502, 857–75, 1913; reprinted in *Niels Bohr: Collected Works*, gen. ed. L. Rosenfeld, Vol. II, Work on Atomic Physics (1912–17), ed. U. Hoyer, North-Holland, Amsterdam, 1981, pp. 159–233. See also J.L. Heilbron & T.S. Kuhn, 'The Genesis of the Bohr Atom', *Hist. Stud. Phys. Sci.*, 1, 211–90, 1969; T. Hirosige, 'The Genesis of the Bohr Atom Model and Planck's Theory of Radiation', *Japanese Studies in History of Science*, 9, 35–47, 1970; J. Mehra & H. Rechenberg, *The*

Historical Development of Quantum theory, Vol. I, Springer-Verlag, New York, 1982, Chs. 2, 3. See also *Coll. Works*, II, pp. 103–134.

3 Letter of Einstein to Bohr, 22 September 1913, *Coll. Works*, II, p. 532.

4 *Albert Einstein: Philosopher-Scientist*, ed. P. A. Schilpp, Northwestern University Press, Evanston, 1949, p. 47.

5 See Bohr's letter of 3 March 1914 to C.W. Oseen, *Coll. Works*, I, p. 128.

6 M. Hesse, *Models and Analogies in Science*, Sheed & Ward, London, 1963, pp. 9ff.

7 'Om Vekselvirkning mellem Lys og stof' ('On the Interaction between Light and Matter'), *Niels Bohr: Collected Works*, gen. ed. L. Rosenfeld, Vol. III, The Correspondence Principle (1918–23), ed. J. Rud Nielsen, North-Holland, Amsterdam, 1976, pp. 229, 234.

8 Bohr, 'The Structure of the Atom', *Nature*, 112, 1923, p. 4.

9 See Bohr, 'On the Application of the Quantum Theory to Atomic Structure', *Proc. Cam. Phil. Soc. (Suppl.)*, 1924, p. 34.

10 Ibid.

11 See Bohr, 'On the Application of the Quantum Theory', p. 35.

12 See AHQP interview with O. Klein, 15 July 1963, session 6 (p. 1).

13 See Bohr, 'On the Application of the Quantum Theory', p. 38. The words 'these phenomena' refer to interference effects.

14 'Silliman Lectures', autumn 1923, *Coll. Works*, III, p. 587.

15 Letter of 21 January 1923, NBA: BSC 16, 2 (cited and translated by R.H. Stuewer, *The Compton Effect*, p. 241).

16 See AHQP interview with Bohr, 15 February 1963 (p. 15).

17 See Bohr, typescript of 1921, 'On the Application of the Quantum Theory to Atomic Problems', *Coll. Works*, III, pp. 374f.; cf. p. 413.

18 See C.G. Darwin's letter of 20 July 1919 to Bohr, *Niels Bohr: Collected Works*, gen. ed. E. Rüdinger, Vol. V, The Emergence of Quantum Mechanics (Mainly 1924–26), ed. K. Stolzenburg, North-Holland, Amsterdam, 1984, p. 314. See also M.J. Klein, 'The First Phase of the Bohr–Einstein Dialogue', *Hist. Stud. Phys. Sci.*, 2, 1970, p. 20.

19 See Einstein's letter of 4 November 1910 to J. Laub, in C. Seelig, *Albert Einstein*, p. 197.

20 See C.G. Darwin, 'A Quantum Theory of Optical Dispersion', *Nature*, 110, 841–2, 1922.

21 See Bohr, 'On the Application of the Quantum Theory to Atomic Structure', *Proc. Cam. Phil. Soc. (Suppl.)*, 1924, p. 39fn.

22 See J.C. Slater, 'Radiation and Atoms', *Nature*, 113, 307–8, 1924.

23 Bohr, 'On the Application of the Quantum Theory', p. 39.

24 Bohr, 'On the Application of the Quantum Theory', p. 36.

25 See J.C. Slater, 'Radiation and Atoms', p. 308.

26 Letter of 10 January 1925, *Coll. Works*, V, p. 67.

27 See letter of 22 July 1924 to D. van Vleck, *Coll. Works*, V, p. 20; see M.J. Klein, 'The First Phase', p. 24.

28 See B.L. van der Waerden (ed.), *Sources of Quantum Mechanics*, North-Holland, Amsterdam, 1967, p. 13.

29 See Bohr, Kramers & Slater, 'The Quantum Theory of Radiation', *Phil. Mag.*, 47, 1924, pp. 790–3 (article dated January 1924).

30 Bohr, Kramers & Slater, 'The Quantum Theory of Radiation', pp. 799ff.

31 See Bohr, Kramers & Slater, 'The Quantum Theory of Radiation', pp. 787, 792; cf. Bohr's letter to Geiger, p. 29 above.

32 See Bohr, 'Über die Wirkung von Atomen bei Stössen', *Z. Phys.*, 34, 142–57, 1925 (received 30 March 1925), translated in *Coll. Works*, V, pp. 194–206.

33 See Einstein's letter of 31 May 1924 to P. Ehrenfest, Einstein Archive (translated by M.J. Klein, 'The First Phase', pp. 32f.)

34 See letter of 2 October 1924, *Coll. Works*, V, pp. 418–21; also in *Wolfgang Pauli: Wissenschaftlicher Briefwechsel mit Bohr, Einstein, Heisenberg u. a., Band I: 1919–1929*, ed. A. Hermann, K.V. Meyenn, V.F. Weisskopf, Springer-Verlag, New York, 1979, pp. 164f., 174, 233.

35 See Bohr, Kramers & Slater, 'The Quantum Theory of Radiation', p. 799.

36 See W. Bothe & H. Geiger, 'Über das Wesen des Comptoneffekts', *Z. Phys.*, 32, 639–63, 1925; see also M. Jammer, *The Conceptual Development of Quantum Mechanics*, McGraw-Hill, New York, 1966, pp. 185f.

37 See A. H. Compton & A.W. Simon, 'Directed Quanta of scattered X-rays', *Phys. Rev.*, 26, 1925, p. 290fn. 6; See R.H. Stuewer, *The Compton Effect*, p. 301.

38 See A.H. Compton & A.W. Simon, 'Directed Quanta', pp. 298ff.

39 Letter of 21 April 1925, *Coll. Works*, V, p. 82.

40 Letter of 21 April 1925, *Coll. Works*, V, p. 353.

41 *Coll. Works*, V, p. 311. By 'symbolic' Bohr means 'formal'.

42 *Coll. Works*, V, p. 497. Bohr remarks that he had recently had a long conversation with Einstein at Leyden (December 1925).

43 See Einstein's letter of 26 April 1926 to E. Schrödinger, in *Letters on Wave Mechanics: Albert Einstein, Erwin Schrödinger, Max Planck, H.A. Lorentz*, ed. K. Przibram, trans. M.J. Klein, Vision, London, 1967, p. 28.

44 *Coll. Works*, V, p. 499.

45 See Bohr, 'Über die Wirkung von Atomen bei Stössen', *Z. Phys.*, 34, 1925, p. 154.

46 Bohr, 'Über die Wirkung', pp. 155ff.

47 Bohr, 'Atomic Theory and Mechanics', *Nature (Suppl.)*, 116, 1925, p. 846 (received 5 December 1925), reprinted in Bohr, *Atomic Theory and the Description of Nature*, Cambridge University Press, Cambridge, 1934, pp. 25–51 (citation from p. 29), and also in *Coll. Works*, V, pp. 273–80. References below will be to the 1961 reprint of *Atomic Theory* ('*ATDN*').

48 *ATDN*, p. 34.

49 See above, p. 21.

50 *ATDN*, pp. 42, 51.

51 *ATDN*, p. 28.

52 *ATDN*, p. 29.

53 *ATDN*, p. 34; cf. pp. 42, 45, 47, 49, 50.

54 *ATDN*, pp. 42, 51.

Chapter 3
From duality to complementarity

1 *ATDN*, p. 51.

2 See L. de Broglie, 'Ondes et quanta', *Compt. Rend. Acad. Sci.*, **177**, 507–10, 1923; 'Quanta de lumière, diffraction et interférences', *Compt. Rend. Acad. Sci.*, pp. 548–50; 'Les quanta, la théorie cinétique et le principe de Fermat', *Compt. Rend. Acad. Sci.*, pp. 630–2; 'A Tentative Theory of Light Quanta', *Phil. Mag.*, **47**, 446–58, 1924. Cf. M. Jammer, *The Conceptual Development*, pp. 243–7.

3 See L. de Broglie, 'Quanta de lumière, diffraction et interférences', p. 549.

4 See S.N. Bose, 'Planck's Gesetz und Lichtquantenhypothese', *Z. Phys.*, **26**, 171–81, 1924; See also M.J. Klein, 'Einstein and the Wave-particle Duality', *The Natural Philosopher*, **3**, 1964, pp. 26f.

5 See Einstein, 'Die Plancksche Theorie der Strahlung und die Theorie der Spezifischen Wärme', *Ann. Phys.*, **22**, 180–90, 1907.

6 See Einstein, 'Quantentheorie des einatomigen Gases', *Sitz. preuss. Akad. Wiss.*, 1924, pp. 261–67 (published 20 September 1924), and 1925, pp. 3–14 (published 9 February 1925).

7 Einstein, 'Quantentheorie des einatomigen Gases', p. 3 (translated by M.J. Klein, 'Einstein and the Wave-particle Duality', p. 36).

8 Einstein, 'Quantentheorie des einatomigen Gases', p. 9 (M.J. Klein's translation).

9 See W. Elsasser, 'Bemerkungen zur Quantenmechanik freier Elektronen', *Naturw.*, **13**, 711, 1925.

10 See C.J. Davisson & L.H. Germer, 'The Scattering of Electrons by a single Crystal of Nickel', *Nature*, **119**, 558–60, 1927; see also G.P. Thomson, 'Diffraction of Cathode Rays by a thin Film', *Nature*, **119**, 890, 1927.

11 Bohr, 'Über die Wirkung von Atomen bei Stössen', *Z. Phys.*, **34**, 1925, p. 157 (postscript dated July 1925).

12 *ATDN*, p. 43; cf. Bohr, 'Linienspektren und Atombau', *Ann. Phys.*, **71**, 1923, p. 276; reprinted in *Niels Bohr: Collected Works*, Vol. IV, The Periodic System (1920–23), ed. J. Rud Nielsen, North-Holland, Amsterdam, 1977, pp. 598, 646.

13 See W. Pauli, 'Über den Einfluss der Geschwindigkeitsabhängigkeit der Elektronenmasse auf den Zeemaneffekt', *Z. Phys.*, **31**, 373–85, 1925; see also P. Forman, 'The Doublet Riddle and Atomic Physics *circa* 1924', *Isis*, **59**, 156–74, 1968; D. Serwer, '*Unmechanischer Zwang*: Pauli, Heisenberg, and the Rejection of the Mechanical Atom, 1923–1925', *Hist. Stud. Phys. Sci.*, **8**, 189–256, 1977.

14 See S. Goudsmit & G.E. Uhlenbeck, 'Ersetzung der Hypothese vom unmechanischen Zwang durch eine Forderung bezüglich des innern Verhaltens jedes einzelnen Elektrons', *Naturw.*, 13, 953–4, 1925. See also B.L. van der Waerden, 'Exclusion Principle and Spin', *Theoretical Physics in the Twentieth Century*, ed. M. Fierz & V.F. Weisskopf, Interscience, New York, 1960, pp. 199–244.

15 See D. Serwer, '*Unmechanischer Zwang*', pp. 250f. Cf. *Pauli: Wiss. Brief.*, I, p. 11.

16 Bohr, 'Spinning Electrons and the Structure of Spectra', *Nature*, 117, 1926, p. 265 (*Coll. Works*, V, p. 289).

17 Letter of 21 May 1925, *Pauli: Wiss. Brief.*, I, p. 216.

18 See W. Heisenberg, 'Über quantentheoretische Umdeutung kinematischer und mechanischer Beziehungen', *Z. Phys.*, 33, 879–93, 1925 (received 29 July 1925); see M. Jammer, *The Conceptual Development*, pp. 196ff.; E. MacKinnon, 'Heisenberg, Models, and the Rise of Matrix Mechanics', *Stud. Hist. Phys. Sci.*, 8, 137–88, 1977.

19 See E. Schrödinger, 'Quantisierung als Eigenwertproblem', *Ann. Phys.*, 79, 361–76, 489–527; 80, 437–90; 81, 109–39, 1926. See M. Jammer, *The Conceptual Development*, pp. 258ff.

20 See Bohr, 'On the Application of the Quantum Theory to Atomic Structure', *Proc. Cam. Phil. Soc. (Suppl.)*., 1924, p. 42 (written in November 1922); also in *Coll. Works*, III, p. 499.

21 *ATDN*, p. 49.

22 See Bohr, 'On the Constitution of Atoms and Molecules', *Phil. Mag.*, 26, 1913, p. 54.

23 *Coll. Works*, II, p. 584; see K.M. Meyer-Abich, *Korrespondenz, Individualität, und Komplementarität: Eine Studie zur Geistesgeschichte der Quantentheorie in den Beiträgen Niels Bohrs*, Steiner, Wiesbaden, 1965, p. 73.

24 *Z. Phys.*, 2, 1920, p. 427 (*Coll. Works*, III, p. 245f). Cf. K.M. Meyer-Abich, *Korrespondenz*, p. 72.

25 See Bohr, 'The Structure of the Atom', *Nature*, 112, 1922, p. 4 (*Coll. Works*, IV, p. 472).

26 Bohr, 'On the Application of the Quantum Theory to Atomic Structure', p. 23 (*Coll. Works*, III, p. 480).

27 See *ATDN*, p. 85.

28 See K.R. Popper, *Conjectures and Refutations*, Routledge & Kegan Paul, London, 4th edn 1972 (orig. 1963), pp. 217ff., 232; See also H.R. Post, 'Correspondence, Invariance and Heuristics', *Stud. Hist. Phys. Sci.*, 2, 1971, pp. 228–35. Post makes no mention of Bohr's view.

29 The notion of idealisation has been stressed by Polish philosophers; see e.g. W. Krajewski, *Correspondence Principle and Growth of Science*, Reidel, Dordrecht, 1977, p. 43.

30 *ATDN*, p. 5.

31 See *ATDN*, pp. 55, 98.

32 H. Høffding, *A History of Modern Philosophy*, trans. B.E. Meyer, MacMillan, London, 1908, Vol. II, p. 58.

33 See E. Schrödinger, 'Quantisierung als Eigenwertproblem', *Ann. Phys.*, 79, 1926, pp. 375, 506; see M. Jammer, *The Philosophy of Quantum Mechanics: Interpretations of Quantum Mechanics in Historical Perspective*, Wiley, New York, 1974, pp. 24ff.

34 See *ATDN*, pp. 74ff.; see M. Born, 'Zur Quantenmechanik der Stossvorgänge', *Z. Phys.*, 37, 863–7; 38, 803–27, 1926 (published 10 July & 14 September respectively); see also M. Jammer, *The Conceptual Development*, pp. 248ff.

35 *ATDN*, p. 75.

36 Typescript entitled 'The Quantum Postulate and the Recent Development of Atomic Theory', 12 October 1927, *Niels Bohr: Collected Works*, gen. ed. E. Rüdinger, Vol. VI, The Foundations of Quantum Physics I (1926–32), ed. J. Kalckar, North-Holland, Amsterdam, 1985, p. 97.

37 *ATDN*, p. 79.

38 Letter of 23 October 1926 (in German), *Coll. Works*, VI, 119, pp. 459f.

39 In the brief notice of the lecture in *Nature*, 119, 1927, p. 26, Bohr states that wave mechanics does not elide discontinuity.

40 *Correspondance entre Harald Høffding et Émile Meyerson*, ed. H. Høffding, Munksgaard, Copenhagen, 1939, p. 131 (cited by M. Jammer, *The Conceptual Development, p. 347*).

41 NBA: BSC 9. Most of the Bohr Archive material cited in this chapter was first presented in K. Stolzenburg, *Die Entwicklung des Bohrschen Komplementaritätgedankens in den Jahren 1924 bis 1929*, doctoral dissertation, University of Stuttgart, 1977.

42 NBA: BSC 15, 1; see Bohr's letter of 2 December 1926 to Schrödinger, *Coll. Works*, VI, p. 14.

43 *ATDN*, p. 91.

44 See W. Heisenberg, 'Über den anschaulichen Inhalt der quantentheoretischen Kinematik und Dynamik', *Z. Phys.*, 43, 1927, pp. 175, 178f, 180. Cf. W. Heisenberg, *The Physical Principles of the Quantum Theory*, trans. F.C. Hoyt, Dover, Chicago, 1930, pp. 16ff.

45 *Pauli: Wiss. Brief.*, I, p. 347.

46 *Pauli: Wiss. Brief.*, I, p. 350.

47 NBA: BSC 11, 2.

48 W. Heisenberg, 'Über den anschaulichen Inhalt', p. 175.

49 W. Heisenberg, 'The Quantum Theory and its Interpretation', *Niels Bohr: His Life and Work as seen by his Friends and Colleagues*, ed. S. Rozental, North-Holland, Amsterdam, 1967, p. 175.

50 W. Heisenberg. 'The Quantum Theory and its Interpretation', p. 101f.

51 NBA: BSC 9. See also W. Heisenberg, 'Die Entwicklung der Quantentheorie 1918–1928, *Naturw.*, 17, 490–6, 1929.

52 See Bohr, *ATDN*, pp. 63ff.

53 Ibid.

54 See *ATDN*, p. 57f.

55 See W. Heisenberg, 'The Quantum Theory and its Interpretation', p. 106.

56 *Coll. Works*, VI, p. 18.

57 W. Heisenberg, 'Über den anschaulichen Inhalt', p. 197.

58 See Bohr's letters to H. Geiger, M. Born, J.C. Slater, cited above, pp. 29f.

59 See Bohr, 'Uber die Wirkung von Atomen bei Stössen', *Z. Phys.*, 34, 1925, pp. 155ff. (cited above, p. 31).

60 Manuscript notes in Danish, NBA: BMSS 12, 1 ('Ubestimthedsrelationens som direkte Konsekvens af Bølge-og Partikel Dilemmat').

61 *ATDN*, p. 114

62 *Coll. Works*, VI, p. 419.

63 *Coll. Works*, VI, p. 420.

64 *Coll. Works*, VI, p. 421.

65 See W. Heisenberg, 'The Development of the Interpretation of the Quantum Theory', *Niels Bohr and the Development of Physics*, ed. W. Pauli, Pergamon, London, 1955, p. 15.

66 AHQP interview, 15 July 1963, p. 22.

67 NBA: BSC 10.

68 NBA: BSC 11, 2.

69 NBA: BSC 14. See K. Stolzenburg, *Die Entwicklung des Bohrschen Komplementaritätgedankens*, p. 150. Cf. W. Pauli's letter of 6 August 1927 to Bohr, *Pauli: Wiss. Brief.*, I, pp. 404ff.

70 See the notes of N.R. Campbell & P. Jordan, 'Philosophical Foundations of Quantum Theory', *Nature*, 119, 1927, p. 779 (*Coll. Works*, VI, p. 58).

71 See Bohr, 'The Quantum Postulate and the Recent Development of Atomic Theory', *Atti del Congresso dei Fisici, Como, 11–20 Settembre 1927*, Zanicelli, Bologna, 1928, Vol. 2, pp. 565–88. An expanded version of this lecture was published in *Nature*, 121, 580–90, 1928; this version is reprinted in *ATDN*, pp. 52–91. It is the *ATDN* version which is cited in the present work. For a collation of the different versions, see *Coll. Works*, VI, pp. 110–2.

72 Undated MS of 8 pp. in O. Klein's handwriting, entitled 'Philosophical foundations of the quantum theory', NBA: BMSS 11, 4 (*Coll. Works*, VI, p. 69). The emendations are in Bohr's own hand.

73 NBA: BMSS 11, 4.

74 *ATDN*, p. 56; cf. pp. 86, 89.

Chapter 4
The meaning of complementarity

1 *APHK*, p. 16; cf. p. 88.

2 *ATDN*, p. 56.

3 *ATDN*, p. 54; cf. p. 19.

4 Bohr, *Essays 1958–1962 on Atomic Physics and Human Knowledge*, Interscience, New York, 1963, p. 92; cf. pp. 4, 60 (abbrev. *Essays*); *APHK*, p. 26.

5 *ATDN*, p. 53.

6 See *ATDN*, p. 54.

7 *Shorter Oxford English Dictionary.*

8 *ATDN*, p. 10.

9 *ATDN*, p. 19.

10 See *ATDN*, pp. 57, 60, 95.

11 *APHK*, p. 74; see also pp. 40, 90, 99; *Essays*, pp. 19, 25, 60, 92.

12 *ATDN*, p. 10.

13 *ATDN*, p. 56.

14 *APHK*, pp. 5, 19.

15 *ATDN*, p. 54.

16 *APHK*, p. 30.

17 Bohr, 'Causality and Complementarity', *Phil. Sci.*, 4, 1937, p. 291.

18 *APHK*, p. 40; see also *Essays*, pp. 4, 12, 92.

19 *APHK*, p. 74; cf. pp. 41, 45, 47, 90; *Essays*, pp. 19, 25, 60.

20 W. Pauli, 'Die Allgemeinen Prinzipien der Wellenmechanik', *Handbuch der Physik*, ed. H. Geiger & K. Scheel, Springer-Verlag, Berlin, 2nd edn 1933, Vol. 24, p. 89 (also in *Pauli: Collected Scientific Papers*, ed. R. Kronig & V.F. Weisskopf, Interscience, New York, 1964, Vol. I, p. 777).

21 *ATDN*, p. 58.

22 See *ATDN*, p. 61; see also Bohr's letter to Einstein cited on p. 52 above.

23 *ATDN*, p. 66; cf. p. 61.

24 Typescript of 56 pp. of lecture notes in Danish entitled 'Kvanteteorien og den klassiske fysiske Teorier', delivered to the Physical Society of Copenhagen, 10 & 24 February, and 24 March 1930; NBA: BMSS 12, 2 (citation from p. 10 of typescript).

25 Bohr, 'The Causality Problem in Atomic Physics', *New Theories in Physics*, International Institute of International Co-operation of the League of Nations, Paris, 1939, p. 18 ('(5)' here refers to the uncertainty relations); cf. *ATDN*, p. 114.

26 See *ATDN*, p. 82.

27 Bohr, 'The Quantum Postulate and the recent Development of Atomic Theory', 12 pp. typescript in English dated 12 & 13 October 1927, *Coll. Works*, VI, p. 98. In a letter (in Danish) of 1 July 1927 to W. Pauli, Bohr states that each use of the notion of individuality limits the use of the superposition principle owing to the loss of the phase accompanying each observation (NBA: BSC 14).

28 *APHK*, p. 30.

29 See M. Born, *The Natural Philosophy of Cause and Chance*, Oxford University Press, Oxford, 1949, p. 105.

30 See L. Rosenfeld, 'Men and Ideas in the History of Atomic Radiation in Space and Time', *Arch. Hist. Ex. Sci.*, 7, 1971, p. 88.

31 See M. Born, *Atomic Physics*, trans. J. Dougall, Blackie, London, 4th edn 1947, p. 92.

32 *ATDN*, p. 55.

33 *APHK*, p. 40. '(3)' here refers to the relation $\Delta p \; \Delta q = h$.

34 *ATDN*, p. 17.

35 *ATDN*, p. 107.

36 From an 8 pp. MS in O. Klein's hand, entitled 'Philosophical Foundations of the Quantum Theory', NBA: BMSS 11, 4 (citations from pp. 2, 4).

37 *ATDN*, p. 56.

38 See *ATDN*, pp. 71ff.

39 *ATDN*, p. 79; cf. p. 111.

40 *ATDN*, pp. 76f.

41 From a 1 p. MS in Danish dated 3 August 1927 (in O. Klein's hand), NBA: BMSS 11, 4.

42 Letter of 9 May 1928 to C. Darwin, NBA: BSC 9, 4.

43 Letter of 14 June 1928 to C. Møller (in Danish), NBA: BSC 14.

44 Unpublished notes in Danish, 1929, NBA: BMSS 12, 1.

45 *ATDN*, p. 111.

46 Bohr, 'Chemistry and the Quantum Theory of Atomic Constitution', *J. Chem. Soc.*, 134, 1932, p. 370 (the word 'ambiguous' here has the usual Bohrian sense of 'not-well-defined'); cf. p. 374.

47 Bohr, 'Maxwell and Modern Theoretical Physics', *Nature*, 128, 1931, p. 691.

48 L. Rosenfeld, 'The Wave-particle Dilemma', *The Physicist's Conception of Nature*, ed. J. Mehra, Reidel, Dordrecht, 1973, p. 252.

49 See AHQP interview with O. Klein, session 6, p. 7.

50 See *ATDN*, pp. 56f; cf. pp. 66, 69, 77.

51 See *ATDN*, pp. 57, 77.

52 *ATDN*, p. 90; cf. p. 12.

53 *ATDN*, p. 115; cf. p. 108.

54 *ATDN*, p. 77.

55 *ATDN*, p. 93; cf. pp. 5, 51, 93, 96, 103, 108, 111.

56 *ATDN*, pp. 116f. The phrase 'mode of perception' is a rendering of the Danish *Anskuelsesform*.

57 See H. Høffding, *A History of Modern Philosophy*, trans. B.E. Meyer, MacMillan, London, 1908, Vol. II, pp. 52, 58.

58 H. Høffding, *A History*, Vol. II, pp. 56f; cf. Høffding, *The Problems of Philosophy*, trans. G.M. Fisher, MacMillan, New York, 1905, pp. 8, 60, 75. See also I. Kant, *The Critique of Pure Reason*, trans. N. Kemp Smith, MacMillan, London, 1929, pp. 231, 248f. (A 208/B 253; A 228/B 280f.).

59 See Bohr's letter of 2 December 1926 to E. Schrödinger, *Coll. Works*, VI, p. 462. Cf. Bohr's letters to C. Darwin (24 November 1926) and to B. Ray (22 January 1927) cited above, p. 45.

60 See H. Høffding, *A History*, Vol. II, p. 58.

61 *ATDN*, p. 5; cf. pp. 55, 98.

62 Ibid.

63 *ATDN*, p. 16.

64 *ATDN*, p. 111.

65 See J. Turner, 'Maxwell on the Method of Physical Analogy', *Brit. J. Phil. Sci.*, 6, 226–38, 1955; see also M. Hesse, *Models and Analogies in Science*, Sheed & Ward, London, 1963.

66 See Bohr, 'Atomic Structure', *Nature*, 107, 1921, pp. 103f; see also N.R. Campbell, 'Atomic Structure', *Nature*, 106, 1920, pp. 408f.

67 H. Høffding, *The Problems of Philosophy*, pp. 31, 122; cf. Høffding, 'On Analogy and its Philosophical Importance', *Mind*, 14, 199–209, 1905.

68 Letter of 22 September 1922 to Høffding (in Danish), NBA: BSC 3, 4.

69 See W. Pauli, 'Niels Bohr on his Sixtieth Birthday', *Revs. Mod. Phys.*, 17, 1945, p. 99.

70 See W. Pauli's letter of 21 February 1924 to Bohr, *Coll. Works*, V, p. 413 (also in *Pauli: Wiss. Brief.*, I, p. 148). See also D. Serwer, '*Unmechanischer Zwang*: Pauli, Heisenberg, and the Reception of the Mechanical Atom, 1923–1925', *Hist. Stud. Phys. Sci.*, 8, 1977, p. 230.

71 *ATDN*, p. 12.

72 See *ATDN*, p. 103; cf. pp. 94, 111, 113.

73 See *ATDN*, pp. 16f; cf. pp. 94, 111, 113.

74 Single MS page in Danish, NBA: BMSS 11, 4.

75 MS of 3 pp. in English and Danish, dated 2 July 1927, NBA: BMSS 11, 4 (citation from p. 3).

76 See *ATDN*, p. 16f; cf. pp. 94, 111, 113.

Chapter 5
The foundations of kinematic–dynamic complementarity

1 *ATDN*, p. 66.

2 See J.L. Park & H. Margenau, 'Simultaneous Measurability in Quantum Theory', *Internat. J. Th. Phys.*, 1, 1968, p. 240.

3 See J.L. Park & H. Margenau, 'Simultaneous Measurability', pp. 244, 263.

4 *ATDN*, p. 62; cf. p. 73. The expression '(2)' refers to the uncertainty relations.

5 See J.L. Park & H. Margenau, 'Simultaneous Measurability', p. 230.

6 Bohr, 'The Causality Problem in Atomic Physics', *New Theories in Physics*, International Institute of International Co-operation of the League of Nations, Paris, 1939, p. 18.

7 See *ATDN*, p. 60.

8 See W. Heisenberg, *The Physical Principles of the Quantum Theory*, trans. F.C. Hoyt, Dover, Chicago, 1930, pp. 39ff.

9 See *ATDN*, p. 66; cf. pp. 98, 114; see also Bohr, 'Causality and Complementarity', *Phil. Sci.*, 4, 1937, p. 292; 'The Causality Problem in Atomic Physics', p. 24; 'On the Notions of Causality and Complementarity', *Dialectica*, 2, 1948, p. 315; *Essays*, pp. 3, 5, 11, 24, 59, 91.

10 See *Essays*, p. 5.
11 See *Essays*, p. 4.
12 *APHK*, p. 72.
13 *ATDN*, p. 53.
14 See *ATDN*, pp. 54f.
15 *Essays*, p. 61; cf. p. 5.
16 See *ATDN*, p. 67; cf. pp. 98, 114.
17 See Bohr, 'Can Quantum-Mechanical Description of Physical Reality be Considered Complete?', *Phys. Rev.*, **48**, 1935, p. 697.
18 For a more detailed analysis see E. Scheibe, *The Logical Analysis of Quantum Mechanics*, trans. J.B. Sykes, Pergamon, Oxford, 1973, pp. 45f.
19 *ATDN*, p. 98; cf. p. 114; see also Bohr, 'Can Quantum-Mechanical Description of Physical Reality be Considered Complete?', p. 698; cf. *Essays*, p. 5.
20 *ATDN*, pp. 11f.
21 See *ATDN*, pp. 13, 54, 75, 80; cf. *APHK*, pp. 73, 89f, 98.
22 See *APHK*, p. 71; *Essays*, pp. 2, 4.
23 See *ATDN*, pp. 7, 53, 108; *APHK*, pp. 6, 17, 19, 34; *Essays*, pp. 24, 60.
24 *APHK*, p. 72; cf. pp. 1, 71, 90; *Essays*, p. 4.
25 See *APHK*, p. 90; cf. pp. 40, 99; see also Bohr, 'Causality and Complementary', *Phil. Sci.*, **4**, 1937, pp. 291, 293; 'The Causality Problem in Atomic Physics', *New Theories in Physics*, pp. 19f; *Essays*, pp. 2, 4, 24, 92.
26 See D. Bohm, *Quantum Theory*, Prentice-Hall, New York, 1951, Ch. 6, Sect. 2.
27 See S-I. Tomonaga, *Quantum Mechanics*, North-Holland, Amsterdam, 1966, Vol. 2, p. 273.
28 See A. Messiah, *Quantum Mechanics*, North-Holland, Amsterdam, 1961, Vol. 1, Ch. 4, Sect. 18.
29 See *APHK*, p. 45; see E. Scheibe, *The Logical Analysis of Quantum Mechanics*, pp. 47f.
30 See *APHK*, p. 49.
31 See *APHK*, pp. 48f; cf. 'Can Quantum-Mechanical Description of Physical Reality be Considered Complete?', *Phys. Rev.*, **48**, 1935, pp. 697f.
32 *APHK*, pp. 46f; see also E. Scheibe, *The Logical Analysis*, pp. 48f.
33 See *ATDN*, pp. 53, 68.
34 *ATDN*, p. 22; cf. pp. 54f.
35 *ATDN*, pp. 11f.
36 *ATDN*, p. 96; cf. p. 15.
37 *Essays*, p. 59; cf. Bohr, 'Causality and Complementarity', *Phil. Sci.*, 1937, p. 293.
38 *ATDN*, p. 54.
39 See *APHK*, p. 33; cf. p. 71; *Essays*, pp. 2, 91.
40 See *ATDN*, pp. 54, 67; cf. E. Scheibe, *The Logical Analysis*, p. 25.
41 See *ATDN*, pp. 15, 96, 99, 119; *APHK*, pp. 80f, 91, 101; *Essays*, pp. 12, 14.
42 See Bohr, 'The Causality Problem in Atomic Physics', *New Theories in Physics*, pp. 24, 28.

43 *ATDN.* p. 67. '(2)' here refers to the uncertainty relations.

44 Bohr, 'Causality and Complementarity', *Phil. Sci.*, **4**, 1937, p. 290.

45 See *APHK*, p. 39; cf. pp. 26, 72, 88f; *ATDN*, pp. 5, 8, 16f, 53, 94; *Essays*, pp. 3, 11, 24, 59, 78, 91.

46 Bohr, 'Can Quantum-Mechanical Description of Physical Reality be Considered Complete?', *Phys. Rev.*, **48**, 1935, p. 701.

47 *Essays*, pp. 3f; cf. pp. 6, 78; *APHK*, pp. 25, 50, 55, 89.

48 *ATDN*, p. 17.

49 *Essays*, p. 3; cf. pp. 24, 61, 92; *APHK*, pp. 51, 73, 88, 90, 98.

50 See Bohr's letter of 16 May 1947 to W. Pauli, *Coll. Works*, VI, pp. 451–54. See I. Kant, *Critique of Pure Reason*, trans. N. Kemp Smith, MacMillan, London, 1929, p. 220 (A 191ff./ B 236ff.).

51 *APHK*, p. 39; cf. pp. 26, 72, 88f; *ATDN*, pp. 5, 8, 16f, 53, 94; *Essays*, pp. 3, 11, 24, 59, 78, 91.

52 *APHK*, p. 26; cf. p. 72; *Essays*, pp. 1, 3, 11, 24, 59, 78, 91.

53 *Essays*, p. 1.

54 *APHK*, p. 72; cf. pp. 26, 39, 88f; *Essays*, pp. 1, 3, 11, 24, 91.

55 *ATDN*, p. 8; cf. pp. 5, 8, 53.

56 *ATDN*, pp. 16f.

57 See *ATDN*, p. 8.

58 Bohr, 'On the Application of the Quantum Theory to Atomic Structure. Part I: the Fundamental Postulates', *Proc. Cam. Phil. Soc. (Suppl.)*, 1924, p. 1; *Coll. Works*, III, p. 458 (This paper was written in 1922.)

59 *Coll. Works*, VI, p. 464. Schrödinger is referring here to Bohr's Como article.

60 Letter of 23 May 1928 to Schrödinger, *Coll. Works*, VI, p. 465.

61 Einstein's letter 31 May 1928 to Schrödinger, *Letters on Wave Mechanics: Albert Einstein, Erwin Schrödinger, Max Planck, H.A. Lorentz*, ed. K. Przibram, trans. M.J. Klein, Vision Press, London, 1967, p. 31. (This work gives the wrong date for this letter.)

62 See *ATDN*, pp. 94f; cf. pp. 1, 8, 53.

63 *ATDN*, p. 5; cf. *Essays*, pp. 5, 11.

64 *APHK*, pp. 65, 67.

65 *APHK*, p. 68.

66 *Essays*, p. 59.

67 *APHK*, p. 67.

68 See P. Feyerabend, 'Complementarity', *Proc. Arist. Soc. (Suppl. Vol.)*, **32**, 1958, pp. 83–86.

69 See M. Bunge, 'Strife about Complementarity', *Brit. J. Phil. Sci.*, **6**, 1955, pp. 2f., 8.

70 'On the Notions of Causality and Complementarity', *Dialectica*, **2**, 1948, p. 317; cf. *ATDN*, p. 17; *APHK*, pp. 50, 57, 64; *Essays*, pp. 4ff, 11, 92.

71 *APHK*, p. 73; cf. p. 64.

72 See *Essays*, p. 3.

73 *Essays*, p. 3; cf. pp. 7, 10, 24; *APHK*, pp. 51, 73, 88, 90, 98.

74 See *APHK*, pp. 64, 73.

75 *ATDN*, p. 97.

76 *ATDN*, p. 115.

77 *APHK*, p. 74.

78 See Bohr's letter of 2 March 1955 to Pauli, in reply to letter of Pauli to Bohr of 15 February 1955, NBA: BSC (cited by H.J. Folse, 'Kantian Aspects of Complementarity', *Kantstudien*, **69**, 1978, p. 64).

79 See Bohr's letter of 25 March (H.J. Folse, 'Kantian Aspects', p. 65).

80 *APHK*, p. 51.

81 Bohr, 'The Causality Problem in Atomic Physics', *New Theories in Physics*, p. 19.

82 *ATDN*, p. 97; cf. p. 53.

83 *Essays*, p. 3.

Chapter 6
Bohr's theory of measurement

1 Bohr, 'Causality and Complementarity', *Phil. Sci.*, **4**, 1937, p. 293.

2 See A. Fine, 'Realism in Quantum Measurements', *Methodology and Science*, **1**, 1968, pp. 215ff.

3 See A. Fine, 'Realism in Quantum Measurement', pp. 218f. I am indebted to Fine here.

4 See A. Fine, 'The Insolubility of the Quantum Measurement Problem', *Phys. Rev.*, **2D**, 2783–87, 1970.

5 See A. Daneri, A. Loinger & G.M. Prosperi, 'Quantum Theory of Measurement and Ergodicity Conditions', *Nuclear Physics*, **33**, 297–319, 1962; see also A. Daneri, A. Loinger & G.M. Prosperi, 'Further Remarks on the Relations between Statistical Mechanics and Quantum Theory of Measurement', *Nuovo Cimento*, **44B**, 119–28, 1966. (A mixture is a weighted sum of homogeneous states.)

6 See W. Weidlich, 'Problems of the Quantum Theory of Measurement', *Z. Phys.*, **205**, 119–220, 1967. For a critique of Weidlich's theory see A. Fine, 'The Insolubility of the Quantum Measurement Problem', *Phys. Rev.*, **2D**, 1970, pp. 2785f.

7 *APHK*, p. 89; cf. pp. 5, 51, 73, 89f, 98; *Essays*, pp. 3, 61, 92.

8 Bohr, 'On the Notions of Causality and Complementarity', *Dialectica*, **2**, 1948, p. 317.

9 See J. Bub, 'The Daneri-Loinger-Prosperi Theory of Measurement', *Nuovo Cimento*, **57B**, 1968, pp. 505, 507.

10 See J.M. Jauch, E.P. Wigner & M.M. Yanase, 'The Daneri-Loinger-Prosperi Theory of Measurement', *Nuovo Cimento*, **57B**, 503–20, 1968.

11 See A. Daneri, A. Loinger & G.M. Prosperi, 'Quantum Theory of Measurement and Ergodicity Conditions', p. 298fn.

12 L. Rosenfeld, 'The Measuring Process in Quantum Mechanics', *Progress of Theoretical Physics*, extra nr., 1965, p. 225 (also in *Selected Papers of Léon Rosenfeld*, ed. R.S. Cohen & J.J. Stachel, Reidel, Dordrecht, 1979, p. 539).

13 L. Rosenfeld, 'The Measuring Process in Quantum Mechanics', p. 230 (*Selected Papers*, p. 546).

14 See C. George, I. Prigogine & L. Rosenfeld, 'The Macroscopic Level of Quantum Mechanics', *Nature*, 240, 25–27, 1972 (also in L. Rosenfeld, *Selected Papers*, pp. 571–98).

15 See L. Rosenfeld, 'The Measuring Process in Quantum Mechanics', p. 580.

16 See *ATDN*, p. 17.

17 See P. Teller, 'Quantum Mechanics and the Nature of Continuous Physical Quantities', *J. Phil.*, 76, 345–61, 1979.

18 See P. Teller, 'The Projection Postulate and Bohr's Interpretation of Quantum Mechanics', *PSA 1980: Proceedings of the 1980 Biennial Meeting of the Philosophy of Science Association*, ed. P.D. Asquith & R.N. Giere, Philosophy of Science Association, East Lansing, Michigan, 1981, pp. 206f, 212, 215f.

19 *Essays*, p. 5.

20 See *Essays*, p. 12; cf. pp. 3ff., 60; *ATDN*, p. 12; *APHK*, pp. 64, 71, 90.

21 *ATDN*, p. 23.

22 See K.R. Popper, *The Logic of Scientific Discovery*, Ch. 9, Sect. 75.

23 See H.P. Robertson, 'The Uncertainty Principle', *Phys. Rev.*, 34, 163–4, 1929.

24 See E. Scheibe, *The Logical Analysis of Quantum Mechanics*, p. 110

25 See P. Busch and P.J. Lahti, 'A Note on Quantum Theory, Complementarity, and Uncertainty', *Phil. Sci.*, 32, 1985, pp. 809–15.

26 See J. von Neumann, *Mathematical Foundations of Quantum Mechanics*, trans. R.T. Meyer, Princeton University Press, Princeton, 1955 (1st edn Springer-Verlag, Berlin, 1933), pp. 351–7, 418ff.

27 See G. Lüders, 'Über die Zustandsänderung durch den Messprozess', *Ann Phys.*, 8, 322–28, 1951.

28 See J. von Neumann, *Mathematical Foundations*, p. 420.

29 See P. Teller, 'The Projection Postulate as a Fortuitous Approximation', *Phil. Sci.*, 50, 1983, p. 427.

30 This terminology is due to W. Pauli (though the distinction is due originally to L.D. Landau & R. Peierls); see W. Pauli's *Collected Scientific Papers*, Vol. 2, p. 843.

31 See B. d'Espagnat, *Conceptual Foundations of Quantum Mechanics*, Benjamin, Reading, Mass., 2nd edn 1976, p. 208.

32 See L.D. Landau & R. Peierls, 'Erweiterung des Ubestimmtheitsprinzips für relativistische Quantentheorie', *Z. Phys.*, 69, 56–69, 1931.

33 See F.W. London & E. Bauer, *La théorie de l'observation en mécanique quantique*, Hermann, Paris, 1939, p. 42. The above is a reconstruction of London and Bauer's argument.

34 See E. Schrödinger, 'Die gegenwärtige Situation in der Quantenmechanik', *Naturw.*, 23, 1935, p. 812.

35 See *APHK*, pp. 68–70; *Essays*, pp. 3, 7, 10, 60.

36 See Bohr. 'The Causality Problem in Atomic Physics', *New Theories in Physics*, p. 24; cf. *APHK*, pp. 64, 73; *Essays*, p. 3.

37 Bohr, 'The Causality Problem in Atomic Physics', *New Theories in Physics*, p. 20.

Chapter 7
Bohr's theory of properties

1 See M. Jammer, *The Philosophy of Quantum Mechanics*, Wiley, New York, 1974, pp. 160ff.

2 N. Bohr & L. Rosenfeld, 'Zur Frage der Messbarkeit der elektromagnetischen Feldgrössen', *Kongelige Danske Videnskabernes Selskabs Mat. Fys. Meddelelelser*, **12**, 1933; translated in *Selected Papers of Léon Rosenfeld* (citation, p. 387f.).

3 Bohr, 'The Causality Problem in Atomic Physics', *New Theories in Physics*, p. 24.

4 *APHK*, p. 26; cf. 74.

5 Bohr, 'The Causality Problem in Atomic Physics', p. 25.

6 *APHK*, p. 25; cf. pp. 39, 47, 52, 98; Bohr, 'Causality and Complementarity', *Phil. Sci.*, **4**, 1937, pp. 290f, 293; 'The Causality Problem in Atomic Physics', p. 193; Bohr, 'On the Notions of Causality and Complementarity', pp. 313, 317.

7 Notes for 'The Causality Problem in Atomic Physics', dated 9 September 1938, NBA: BMSS 15, 2.

8 See J.L. Park & H. Margenau, 'Simultaneous Measurability in Quantum Theory', *Internat. J. Th. Phys.*, **1**, 1968, pp. 217, 218; cf. Margenau, *The Nature of Physical Reality*, McGraw-Hill, New York, 1950, p. 326f.

9 See J.L. Park & H. Margenau, 'Simultaneous Measurability', p. 218.

10 W. Heisenberg, *Physics and Philosophy*, Allen & Unwin, London, 1959, p. 160; cf. 40.

11 J. Bub, *The Interpretation of Quantum Mechanics*, Reidel, Dordrecht, 1974, p. 43.

12 See J. Bub, *The Interpretation*, pp. 44f, 128.

13 P. Feyerabend, 'Problems of Microphysics', *Frontiers of Science and Philosophy*, ed. R.G. Colodny, Allen & Unwin, London, 1962, pp. 217ff.

14 See L. Rosenfeld, 'Misunderstandings about the Foundations of Quantum Theory', *Observation and Interpretation*, ed. S. Körner, Butterworths, London, 1957, p. 42.

15 See P. Feyerabend, 'Problems of Microphysics', p. 219.

16 Bohr, 'The Causality Problem in Atomic Physics', p. 20; cf. *APHK*, pp. 50, 57, 90; *Essays*, pp. 4, 92.

17 See *APHK*, p. 64.

18 *APHK*, p. 98; cf. pp. 40, 51, 61.

19 *ATDN*, p. 18.

20 Bohr, 'Causality and Complementarity', *Phil. Sci.*, **4**, 1937, p. 293; cf. *APHK*, p. 73.

21 See M. Bunge, 'Strife about Complementarity', *Brit. J. Phil. Sci.*, **6**, 1–12, 141–54, 1955.

22 Letter in German, NBA: BSC 19, 3. The papers referred to are the 'EPR papers' of Einstein and Bohr.

23 See A. Grünbaum, 'Complementarity in Quantum Physics and its Philosophical Generalisation', *J. Phil.*, **54**, 1957, pp. 716f.

24 See *ATDN*, pp. 59f., 73 respectively; cf. pp. 55, 63, 66, 77, 78, 87.

25 See *ATDN*, pp. 54, 57, 60, 67, 77, 78.

26 See A. Grünbaum, 'Complementarity in Quantum Physics', pp. 713, 717.

27 *ATDN*, p. 61.

28 See A. Grünbaum, 'Complementarity in Quantum Physics', p. 717, fn. 7.

29 See W. Heisenberg, 'Über den anschaulichen Inhalt der quantentheoretischen Kinematik und Dynamik', *Z. Phys.*, **43**, 1927, pp. 879ff.; cf. Heisenberg, *The Physical Principles of the Quantum Theory*, p. 15.

30 *ATDN*, p. 59.

31 *ATDN*, p. 63.

32 Bohr, 'Über die Wirkung von Atomen bei Stössen', *Z. Phys.*, **34**, 1925, pp. 155ff.

33 *ATDN*, p. 55.

34 *ATDN*, p. 79.

35 See A. Grünbaum, 'Complementarity in Quantum Physics', p. 717.

36 See P. Feyerabend, 'Problems of Microphysics', *Frontiers of Science and Philosophy*, ed. R.G. Colodny, University of Pittsburgh Press, Pittsburgh, 1962, pp. 196–198, 259 n. 39.

37 See *ATDN*, p. 81.

38 *ATDN*, p. 54.

39 *ATDN*, p. 11. 'phenomenon' here means simply 'object'.

40 *APHK*, p. 40; cf. pp. 62, 73; 'On the Notions of Causality and Complementarity', *Dialectica*, **2**, 1948, p. 315.

41 Bohr, 'Causality and Complementarity', *Phil. Sci.*, **4**, 1937, pp. 289f; cf. *APHK*, pp. 1, 18, 24, 41, 62, 68, 85, 91, 97, 98; *Essays*, pp. 1, 3, 5, 18. See also A. Petersen, *Quantum Physics and the Philosophical Tradition*, M.I.T. Press, Cambridge, Mass., 1968, pp. 145ff, 168f, 183; see also C.A. Hooker, 'The nature of Quantum Mechanical Reality', *Paradigms and Paradoxes*, ed. R.G. Colodny, University of Pittsburgh Press, Pittsburgh, 1972, pp. 135–9, 146, 167.

42 *APHK*, p. 41.

43 *ATDN*, p. 5; cf. pp. 55, 98, 116.

44 See Bohr, 'The Causality Problem in Atomic Physics', *New Theories in Physics*, pp. 25f; cf. 'Causality and Complementarity', *Phil. Sci.*, **4**, 1937, p. 291; *Essays*, p. 10.

45 *APHK*, p. 24; cf. p. 25.

46 *APHK*, p. 64.

47 See Bohr's letter of 24 August 1927 to Heisenberg, NBA: BSC 11, 2.

48 *Essays*, p. 18; cf. p. 9; *APHK*, pp. 1, 19, 34, 62.

49 See *APHK*, pp. 34, 85.

50 *APHK*, pp. 67, 68.

51 *APHK*, p. 82.

52 Bohr, 'Can Quantum-Mechanical Description of Physical Reality be Considered Complete?', *Phys. Rev.*, **48**, 1935, p. 699.

53 *APHK*, p. 99; cf. p. 92.

54 Bohr, 'The Causality Problem in Atomic Physics', *New Theories in Physics*, p. 20.

55 Bohr, 'The Causality Problem', p. 21.

56 See P. Feyerabend, 'Problems of Microphysics', *Frontiers of Science and Philosophy*, ed. R.G. Colodny, p. 219.

57 Bohr, 'The Causality Problem in Atomic Physics', p. 24.

58 Bohr, 'Can Quantum-Mechanical Description of Physical Reality be Considered Complete?', *Phys. Rev.*, 48, 1935, p. 697.

59 Bohr, 'Can Quantum-Mechanical Description', p. 699; cf. p. 700.

60 See *Essays*, pp. 4, 10f, 18, 24, 91f.

61 See P.F. Strawson, *Introduction to Logical Theory*, Methuen, London, 1952, pp. 175–9.

62 See B. van Fraassen, 'Presupposition, Implication, and Self-reference', *J. Phil.*, 65, 136–52, 1968.

63 See K. Lambert, 'Logical Truth and Microphysics', *The Logical Way of Doing Things*, ed K. Lambert, Yale University Press, New Haven, 1969, pp. 93–117.

64 See B. van Fraassen & C.A. Hooker, 'A Semantic Analysis of Niels Bohr's Philosophy of Quantum Theory', *Foundations of Probability Theory, Statistical Inference, and Statistical Theories of Science*, ed. W.L. Harper & C.A. Hooker, Reidel, Dordrecht, 1976, Vol. 3, pp. 211–41.

65 See Bohr, 'On the Notions of Causality and Complementarity', *Dialectica*, 2, 1948, p. 317; *Essays*, pp. 5f.

66 Bohr, 'The Causality Problem in Atomic Physics', *New Theories in Physics*, p. 25.

Chapter 8
Einstein *versus* Bohr

1 See Einstein's comments, *Électrons et photons: rapports et discussions du cinquième conseil de physique, tenu à Bruxelles du 24 au 29 octobre 1927 sous les auspices de L'Institut International de Physique Solvay*, Gauthier-Villars, Paris, 1928, pp. 254f; see also *APHK*, pp. 41f.

2 See Einstein, 'Électrons et photons', p. 255.

3 Heisenberg states that Einstein was the originator of the statistical interpretation: see W. Heisenberg, 'Über den anschaulichen Inhalt der quantentheoretischen Kinematik und Dynamik', *Z. Phys.*, 43, 1927, p. 176, fn. 1.

4 See *APHK*, p. 44.

5 See *APHK*, p. 45.

6 See *APHK*, p. 46.

7 See *APHK*, pp. 49f.

8 See *APHK*, pp. 52ff.

9 See K.R. Popper, *The Logic of Scientific Discovery*, Hutchinson, London, 9th impression 1977 (original English edn 1959), p. 447.

10 See M. Jammer, *The Philosophy of Quantum Mechanics*, pp. 132ff.

11 See K.R. Popper, *The Logic of Scientific Discovery*, p. 447.

12 See K.R. Popper, *The Logic of Scientific Discovery*, p. 444.

13 See A. Pais, '*Subtle is the Lord . . .': the Science and the Life of Albert Einstein*, Oxford University Press, Oxford, 1982, p. 180f; see also E.M. MacKinnon, *Scientific Explanation and Atomic Physics*, University of Chicago Press, Chicago, pp. 334f. See also *APHK*, p. 54.

14 See W. Heisenberg, 'Über den anschaulichen Inhalt der quantentheoretischen Kinematik und Dynamik', *Z. Phys.*, **43**, 1927, p. 179.

15 See N. Bohr & L. Rosenfeld, 'Zur Frage der Messbarkeit der Eletromagnetischen Feldgrössen', *Kongelige Danske Videnskabernes Selskabs Mat. Fys. Meddelelelser*, **12**, 1–65, 1933; translated in *Selected Papers of Léon Rosenfeld*, p. 374.

16 See *ATDN*, pp. 57ff.; see also M. Jammer, *The Philosophy of Quantum Mechanics*, pp. 136–51.

17 See *APHK*, p. 56.

18 See *APHK*, p. 57.

19 See A. Einstein, R.C. Tolman & B. Podolsky, 'Knowledge of Past and Future in Quantum Mechanics', *Phys. Rev.*, **37**, 780–81, 1931.

20 Letter of 9 July 1931; see M. Jammer, *The Philosophy of Quantum Mechanics*, pp. 171f.

21 *APHK*, p. 56.

22 See M. Jammer, *The Philosophy of Quantum Mechanics*, pp. 171f.

23 See Bohr, 'Can Quantum-Mechanical Description of Physical Reality be Considered Complete?', *Phys. Rev.*, **48**, 1935, p. 698.

24 See J.A. Wheeler, 'The "Past" and the "Delayed Choice" Double-Slit Experiment', *Mathematical Foundations of Quantum Theory*, ed. A.R. Marlow, Academic Press, New York, 1978, pp. 9–48.

25 See W.C. Wickes, C.O. Alley & O. Jakubowicz, 'A "Delayed Choice" Quantum Mechanics Experiment', *Quantum Theory and Measurement*, ed. J.A. Wheeler & W. Zurek, Princeton University Press, Princeton, New Jersey, 1983, pp. 457–61.

26 See *APHK*, p. 57.

27 Bohr, 'The Causality Problem in Atomic Physics', *New Theories in Physics*, p. 21.

28 A. Einstein, B. Podolsky & N. Rosen. 'Can Quantum-Mechanical Description of Physical Reality be Considered Complete?', *Phys. Rev.*, **47**, 777–80, 1935; see also M. Jammer, *The Philosophy of Quantum Mechanics*, pp. 166f. I shall employ in the text the standard abbreviation 'EPR'.

29 See A. Einstein, B. Podolsky & N. Rosen, 'Can Quantum-Mechanical Description', p. 799. I have made a slight alteration to the EPR notation.

30 See Ibid.

31 See D. Bohm, *Quantum Theory*, Prentice-Hall, New York, 1951, pp. 614ff.

32 See C.S. Wu & I. Shaknov, 'The Angular Correlations of Scattered Annihilation Radiation', *Phys. Rev.*, **77**, 136, 1950.

33 A. Einstein, B. Podolsky & N. Rosen, 'Can Quantum-Mechanical Description', p. 777. For a different analysis see L. Wessels, 'The "EPR" Argument: a Post-Mortem', *Philosophical Studies*, **40**, 3–30, 1981; see also J.F. Halpin, 'EPR Resuscitated: a Reply to Wessels', *Philosophical Studies*, **44**, 111–4, 1983.

34 A. Einstein, B. Podolsky & N. Rosen, 'Can Quantum-Mechanical Description',
    p. 777.
35 See A. Einstein, B. Podolsky & N. Rosen, 'Can Quantum-Mechanical Description',
    p. 780.
36 See Einstein, 'Quantenmechanik und Wirklichkeit', *Dialectica*, 2, 1948, p. 323; and
    'Autobiographical Notes', p. 85.
37 See A. Einstein, B. Podolsky & N. Rosen, 'Can Quantum-Mechanical Description',
    p. 779f.
38 See Einstein, 'Physics and Reality', *Journal of the Franklin Institute*, 221, 1936, pp.
    341f. (reprinted in Einstein, *Out of my Later Years*, Thames and Hudson, London,
    1950, p. 90); see also Einstein's letter to Popper in K.R. Popper, *The Logic of
    Scientific Discovery*, 9th impression 1977, p. 459; Einstein, 'Quantenmechanik und
    Wirklichkeit', p. 323; Einstein, 'Reply to Criticisms', *Albert Einstein*, ed P.A.
    Schilpp, pp. 681f.
39 Ibid.
40 See A. Einstein, B. Podolsky & N. Rosen, 'Can Quantum-Mechanical Description',
    pp. 779f.
41 Bohr, 'Quantum Mechanics and Physical Reality', *Nature*, 136, 1935, p. 65.
42 See Bohr, 'Can Quantum-Mechanical Description of Physical Reality be Considered
    Complete?', *Phys. Rev.*, 48, 1935, p. 697.
43 See Bohr, 'Can Quantum-Mechanical Description', pp. 697ff.
44 Bohr, 'Can Quantum-Mechanical Description', p. 699.
45 See Bohr, 'Can Quantum-Mechanical Description', pp. 698f.
46 Ibid; cf. *APHK*, pp. 57, 60.
47 Bohr, 'The Causality Problem in Atomic Physics', *New Theories in Physics*, p. 21.
48 Bohr, 'Can Quantum-Mechanical Description', p. 699.
49 Bohr, 'Can Quantum-Mechanical Description', p. 700.
50 Einstein, *Albert Einstein*, ed. P.A. Schilpp, p. 85.
51 See Bohr, 'On the Notions of Causality and Complementarity', *Dialectica*, 2, 1948,
    p. 316; cf. L. Rosenfeld, 'Niels Bohr in the Thirties', *Niels Bohr*, ed. S. Rozental, p.
    130.
52 See Bohr, 'Causality and Complementarity', *Phil. Sci.*, 4, 1937, p. 293; 'The
    Causality Problem in Atomic Physics', p. 19; *APHK*, p. 40.
53 Bohr, 'Causality and Complementarity', p. 293.
54 See *ATDN*, p. 54; cf. pp. 111, 115; *APHK*, p. 7.
55 *ATDN*, p. 18.
56 See A. Einstein, B. Podolsky & N. Rosen, 'Can Quantum-Mechanical Description',
    p. 780.
57 See Einstein, 'Reply to Criticisms', *Albert Einstein*, ed. P.A. Schilpp, pp. 681, 674.
58 See D. Howard, 'Einstein on Locality and Separability', *Stud. Hist. Phil. Sci.*, 16,
    1985, p. 179, fn. 9.
59 See D. Howard, 'Einstein on Locality', p. 180, fn. 20.
60 See D. Howard, 'Einstein on Locality', p. 178, fn. 17, p. 180.

61 A. Einstein, B. Podolsky & N. Rosen, 'Can Quantum-Mechanical Description', p. 780.

62 See K.R. Popper, *The Logic of Scientific Discovery*, p. 459.

63 See D. Howard, 'Einstein on Locality', p. 183.

64 See Einstein, 'Reply to Criticisms', *Albert Einstein*, ed. P.A. Schilpp, pp. 84, 86; cf. pp. 681f.

65 See Einstein, 'Reply to Criticisms', *Albert Einstein*, ed. P.A. Schilpp, p. 682.

66 See Einstein, 'Reply to Criticisms', *Albert Einstein*, ed. P.A. Schilpp, p. 681.

67 See Bohr, 'The Causality Problem in Atomic Physics', *New Theories in Physics*, p. 20; cf. p. 23.

68 See C.A. Hooker, 'The Nature of Quantum Mechanical Reality: Einstein versus Bohr', *Paradigms and Paradoxes*, ed. R.G. Colodny, University of Pittsburgh Press, Pittsburgh, 1972, p. 256.

Chapter 9
The sequel to the Bohr-Einstein debate

1 See A.M. Gleason, 'Measures on the Closed Subspaces of a Hilbert Space', *Journal of Mathematics and Mechanics*, 6, 885–93, 1957.

2 See D. Bohm, 'A Suggested Interpretation of the Quantum theory in Terms of "Hidden Variables" ', *Phys. Rev.*, 85, 166–79, 180–93, 1952.

3 See J.S. Bell, 'On the Einstein Podolsky Rosen Paradox', *Physics*, 1, 195–200, 1964.

4 See E.P. Wigner, 'On Hidden Variables and Quantum Mechanical Probabilities', *American Journal of Physics*, 38, 1005–9, 1970.

5 See G. Lochak, 'Has Bell's Inequality a General Meaning for Hidden Variables Theories?', *Foundations of Physics*, 6, 1976, esp. pp. 174, 179.

6 See A. Fine, 'How to Count Frequencies: a Primer for Quantum Realists', *Synthèse*, 42, 145–54, 1979; 'Some Local Models for Correlation Experiments', *Synthèse*, 50, 279–94, 1982.

7 See T.K. Lo & A. Shimony, 'Proposed Molecular Test of Local Hidden Variables Theories', *Phys. Rev. A*, 23, 3003–12, 1981.

8 See J. Jarrett, 'On the Physical Significance of the Locality Conditions in the Bell Arguments', *Noûs*, 18, 1984, p. 573. The notation in the present text differs from Jarrett's.

9 See J. Jarrett, 'On the Physical Significance', p. 578.

10 See J. Jarrett, 'On the Physical Significance', p. 582.

11 See J. Jarrett, 'On the Physical Significance', p. 585.

12 See D. Howard, 'Einstein on Locality and Separability', *Stud. Hist. Phil. Sci.*, 16, 1985, pp. 173, 195f.

13 See D. Howard, 'Einstein on Locality', p. 198.

14 See N.C. Petroni & J.-P. Vigier, 'Causal Action-at-a-Distance Interpretation of the Aspect-Rapisarda Experiments', *Physics Letters*, 93A, 383–87, 1983.

15 See M. Redhead, 'Non-locality and Peaceful Co-existence', *Space, Time and Causality*, ed. R. Swinburne, Reidel, Dordrecht, 1983, pp. 151–89.

16 See Einstein, 'Physics and Reality', *Journal of the Franklin Institute*, **221**, 349–82, 1936 (reprinted in Einstein, *Out of my Later Years*, pp. 59–97). References in the text are to the latter edition.

17 Einstein, *Out of my Later Years*, pp. 59f.

18 Ibid. Cf. Einstein, 'Remarks on the Philosophy of Bertrand Russell', *The Philosophy of Bertrand Russell*, ed. P.A. Schilpp, Northwestern University Press, Evanston, 1944, p. 287.

19 See Einstein, 'Remarks on the Philosophy of Bertrand Russell', p. 61; cf. 'Autobiographical Notes', *Albert Einstein*, ed. P.A. Schilpp, p. 13.

20 See Einstein, 'Physics and Reality', p. 60.

21 Einstein, 'Reply to Criticisms', *Albert Einstein*, ed. P.A. Schilpp, p. 699; cf. 673; 'Physics and Reality', pp. 60f.

22 See Einstein, 'Reply to Criticisms', *Albert Einstein*, ed. P.A. Schilpp, p. 674; cf. pp. 678, 680; see also Einstein, *The World as I See It*, trans. A. Harris, Lane, London, 1935, pp. 134f.

23 See Einstein, 'Physics and Reality', pp. 62f.

24 Einstein, 'The Problems of Space, Ether and the Field in Physics', *The World as I See It*, p. 180; cf. 'Autobiographical Notes', p. 13.

25 See Einstein, 'Physics and Reality', p. 29; cf. pp. 36f, 96.

26 See Einstein, 'Physics and Reality', pp. 63f.

27 See Einstein, 'Autobiographical Notes', pp. 27, 37; see also Einstein, 'Considerations concerning the Fundamentals of Physics', *Science*, **91**, 487–92, 1940.

28 See Einstein, *The World as I See It*, pp. 141, 184.

29 Einstein, 'Considerations concerning the Fundamentals of Physics', p. 89; cf. 'Physics and Reality', p. 72.

30 See Einstein, *The World as I See It*, p. 136.

31 See Einstein, *The World as I See It*, pp. 167f; cf. 'Autobiographical Notes', pp. 53, 57.

32 See Einstein, *Ideas and Opinions*, ed. C. Seelig, Redman, London, 1956, pp. 54f.

33 See Einstein, 'Physics and Reality', p. 74.

34 See Einstein, 'Autobiographical Notes', p. 11.

35 Einstein, 'Reply to Criticisms', *Albert Einstein*, ed. P.A. Schilpp, p. 673.

36 W. Whewell, *The Philosophy of the Inductive Sciences: Founded upon their History, a Facsimile of the Second Edition, London, 1847*, ed J. Herivel, Johnson Reprint, New York, 1967, Vol. 1, p. 30.

37 See Einstein, 'Physics and Reality', p. 60; see W. Whewell, *The Philosophy of the Inductive Sciences*, Vol. 1, p. 39.

38 See W. Heisenberg, *Physics and Beyond: Encounters and Conversations*, Allen & Unwin, London, 1971, p. 63.

39 See W. Whewell, *The Philosophy of the Inductive Sciences*, Vol. 2, pp. 36ff.

40 See Einstein, 'Autobiographical Notes', pp. 21f.

41 See W. Whewell, *The Philosophy of the Inductive Sciences*, Vol. 1, pp. 77ff.

42 See Einstein, *Über die Spezielle und die Allgemeine Relativitätstheorie, Gemeinverständlich*, Vieweg, Braunschweig, 1971, p. 5.

43 See Einstein, 'Reply to Criticisms', p. 669.

Chapter 10
Bohr's philosophy of physics

1 For a similar account see W. Newton-Smith, *The Rationality of Science*, Routledge & Kegan Paul, London, 1981. Ch. 2.

2 See W.V.O. Quine, *Word and Object*, M.I.T. Press, Cambridge, Mass., 1960, Ch. 2; *Ontological Relativity and Other Essays*, Columbia University Press, New York, 1969, Ch. 1; see also P. Smith, *Realism and the Progress of Science*, Cambridge University Press, Cambridge, 1981.

3 See H. Putnam, 'Models and Reality', *Journal of Symbolic Logic*, **45**, 464–82, 1980 (also in H. Putnam, *Realism and Reason: Philosophical Papers, Vol. 3*, Cambridge University Press, Cambridge, 1983, Ch. 1).

4 For a recent version of this argument see N. Rescher, *Conceptual Idealism*, Blackwell, Oxford, 1973. There are hints of this argument in Putnam; see H. Putnam, *Realism and Reason*, p. 207.

5 See H. Putnam, *Realism and Reason*, p. 267f.

6 See W. Newton-Smith, *The Rationality of Science*, pp. 41–3; see also H. Putnam, *Meaning and the Moral Sciences*, Routledge & Kegan Paul, London, 1978, pp. 133ff.

7 See L. Bergström, 'Underdetermination and Realism', *Erkenntnis*, **21**, 349–65, 1984.

8 See W. Newton-Smith, *The Rationality of Science*, p. 14; H. Putnam, *Meaning and the Moral Sciences*, pp. 183f.

9 See J.J.C. Smart, *Between Science and Philosophy: an Introduction to the Philosophy of Science*, Random House, New York, 1968, pp. 150f; H. Putnam, *Meaning and the Moral Sciences*, pp. 18f.

10 See R. Boyd, 'On the Current Status of the Issue of Scientific Realism', *Erkenntnis* **19**, 1983, pp. 65f.

11 See G. Oddie, *Likeness to Truth*, Reidel, Dordrecht, 1986 for a thorough treatment of this issue.

12 See L. Laudan, *Progress and its Problems*, Routledge & Kegan Paul, London, 1977. For a critique of realism see L. Laudan, 'A Confutation of Convergent Realism', *Phil. Sci.*, **48**, 19–49, 1981.

13 See B. van Fraassen, *The Scientific Image*, Oxford University Press, Oxford, 1980.

14 See W. Sellars, 'Scientific Realism and Irenic Instrumentalism', *Boston Studies in the Philosophy of Science*, ed. R.S. Cohen & M. Wartofsky, Reidel, Dordrecht, Vol. 2, 1965, p. 189, fn. 23.

15 P.F. Strawson, 'Perception and its Objects', *Perception and Identity: Essays Presented to A.J. Ayer with his Replies to them*, ed. G.F. MacDonald, MacMillan, London, 1979, pp. 58f.

16 See M. Dummett, 'Common Sense and Physics', *Perception and Identity*, ed. G.F. MacDonald, pp. 18–21, 37.

17 See M. Dummett, 'Common Sense and Physics', *Perception and Identity*, ed G.F. MacDonald, pp. 37, 39.

18 See P.F. Strawson, *Individuals: an Essay in Descriptive Metaphysics*, Methuen, London, 1959, pp. 18–21. The affinity between Bohr and Strawson has been noted by E.M. MacKinnon, *Scientific Explanation and Atomic Physics*, University of Chicago Press, Chicago, 1982, pp. 360ff.

19 See P.F. Strawson, *Individuals*, p. 22.

20 See P.F. Strawson, *Individuals*, pp. 38f.

21 See P.F. Strawson, *Individuals*, pp. 44f.

22 *Essays*, p. 78; cf. *APHK*, pp. 26, 39, 88f.

23 See P.F. Strawson, *Individuals*, pp. 34, 81.

24 Letter of Bohr to B. Ray, 22 January 1927, NBA: BSC 16, 2.

25 See P.F. Strawson, *Individuals*, pp. 61f.

26 See P.F. Strawson, *Individuals*, p. 72.

27 See *ATDN*, p. 96.

28 See P.F. Strawson, *Individuals*, p. 62.

29 *ATDN*, p. 91.

30 See P.F. Strawson, 'Perception and its Objects', *Perception and Identity*, ed. G.F. MacDonald, pp. 41–60, esp. 41–7, 57–9.

31 *ATDN*, p. 103; cf. Bohr, 'Chemistry and the Quantum Theory of Atomic Constitution', *J. Chem. Soc.*, **134**, 1932, p. 349.

32 *APHK*, p. 57; see also pp. 40, 64, 71; *Essays*, pp. 3, 12, 60, 92.

33 See *Essays*, p. 5; cf. *ATDN*, p. 12.

34 See M. Dummett, *The Interpretation of Frege's Philosophy*, Duckworth, London, 1981, p. 434; cf. M. Dummett, 'Realism', *Synthèse*, **52**, 1982, p. 55.

35 See M. Dummett, 'Realism', p. 56.

36 See M. Dummett, 'Realism', pp. 69, 101.

37 *Essays*, p. 10.

38 See *APHK*, pp. 26, 72; *Essays*, pp. 1, 3, 11.

39 See Bohr, 'Causality and Complementarity', *Phil. Sci.*, **4**, 1937, p. 292; 'The Causality Problem in Atomic Physics', *New Theories in Physics*, p. 16.

40 See M. Born, *The Born-Einstein Letters*, p. 222f; cf. Einstein, *Albert Einstein*, ed. P.A. Schilpp, pp. 214f, 682f.

41 See *The Born-Einstein Letters*, p. 218.

42 See Einstein, *Albert Einstein*, ed. P.A. Schilpp, p. 699.

43 See P. Teller, 'Quantum Mechanics and the Nature of Continuous Physical Quantities', *J. Phil.*, **76**, 1979, pp. 355ff.

44 *ATDN*, p. 18; cf. pp. 1, 12; *APHK*, pp. 40, 57, 64, 71; *Essays*, pp. 3, 12, 60, 92.

45 See M. Dummett, *Frege: Philosophy of Language*, Duckworth, London, 1973, pp. 467f, 515; M. Dummett, 'What is a Theory of Meaning? (II)', *Truth and Meaning: Essays in Semantics*, ed. G. Evans & J. MacDowell, Oxford University Press, Oxford,

1976, pp. 70, 81f; M. Dummett, *Truth and Other Enigmas*, Duckworth, London, 1978, pp. 16f, 24, 146, 165.

46 See M. Dummett, *Frege: Philosophy of Language*, pp. 586ff, 592; see also M. Dummett, *Truth and Other Enigmas*, pp. 16f.

47 See C.S. Peirce, *Collected Papers of Charles Sanders Peirce*, ed. C. Hartshorne & P. Weiss, Harvard University Press, Cambridge, Mass., 1934, Vol. 5, pp. 402, 427.

48 See C.S. Peirce, *The Collected Papers of Charles Sanders Peirce*, Vol. 5, pp. 403, 412, 483.

49 H. Vaihinger, *The Philosophy of 'As If'*, trans. C.K. Ogden, Kegan Paul, Trench, Trubner, London, 1924, p. 15; cf. pp. 19, 27–32, 96.

50 See H. Høffding, *The Problems of Philosophy*, trans. G.M. Fisher, MacMillan, London, 1906, p. 72.

51 H. Høffding, 'A Philosophical Confession', *The Journal of Philosophy, Psychology and Scientific Methods*, 2, 1905, p. 86f. See also H. Høffding, *The Problems of Philosophy*, pp. 60, 75.

52 See H. Høffding, *The Problems of Philosophy*, p. 8.

53 H. Høffding, 'On Analogy and its Philosophical Importance', *Mind*, 14, 1905, p. 200.

54 See H. Høffding, *The Problems of Philosophy*, pp. 81ff.

55 See H. Høffding, *The Problems of Philosophy*, p. 90.

56 See W. James, *Pragmatism*, ed. H.S. Thayer, Harvard University Press, Cambridge, Mass., 1975 (1st edn 1907), pp. 16, 32, 34, 94, 106, 120.

57 See H. Høffding, *The Problems of Philosophy*, p. 131.

58 See H. Høffding, *The Problems of Philosophy*, pp. 88ff.

59 See H. Høffding, 'A Philosophical Confession', *The Journal of Philosophy, Psychology and Scientific Methods*, 2, 1905, p. 91.

60 H. Høffding, *The Problems of Philosophy*, p. 114; cf. pp. 85, 112f.

61 See H. Høffding, *Modern Philosophers*, trans. A.C. Mason, MacMillan, London, 1915, p. 112.

62 H. Høffding, 'A Philosophical Confession', p. 91.

63 See H. Høffding, *The Problems of Philosophy*, p. 120; cf. H. Høffding, *A History of Modern Philosophy*, trans. B.E. Meyer, MacMillan, London, 1908, Vol. 2, pp. 286ff. See also H. Høffding, 'A Philosophical Confession', p. 90.

64 See Bohr, *Coll. Works*, I, p. 503; *Niels Bohr*, ed. S. Rozental, p. 47; J. Rud Nielsen, 'Memories of Niels Bohr', *Physics Today*, 16, 1963, pp. 27f. See also G. Holton, *Thematic Origins of Scientific Thought: Kepler to Einstein*, Harvard University Press, Cambridge, Mass., 1973, pp. 144f; L.S. Feuer, *Einstein and the Generations of Science*, Basic Books, New York, 1974, pp. 134–40.

65 H. Høffding, 'A Philosophical Confession', p. 91.

66 See H.P. Stapp, 'The Copenhagen Interpretation', *American Journal of Physics*, 40, 1098–1116, 1972.

67 See K.M. Meyer-Abich, *Korrespondenz, Individualität und Komplementarität*, p. 148; cf. pp. 133f., 154; see also W. James, *The Principles of Psychology*, MacMillan, London, 1891, Vol. 1, p. 206.

68 *AHQP* interview, with T.S. Kuhn and A. Petersen, November 1962.

69 See G. Holton, *Thematic Origins*, pp. 137f; and L.S. Feuer, *Einstein and the Generations*, pp. 112–26.

70 See e.g. Bohr, 'The Causality Problem in Atomic Physics', *New Theories in Physics*, p. 25. See also the present text, pp. 72, 95 above.

71 See C.A. Hooker, 'The Nature of Quantum Mechanical Reality: Einstein versus Bohr', *Paradigms and Paradoxes*, ed. R.G. Colodny, University of Pittsburgh Press, Pittsburgh, 1972, pp. 135–8, 156, 167–172; J. Honner, 'The Transcendental Philosophy of Niels Bohr', *Stud. Hist. Phil. Sci.*, 13, 1–29, 1982.

72 I. Kant, *The Critique of Pure Reason*, trans. N. Kemp Smith, MacMillan, London, 1929, p. 259 (B 298/A 239).

73 See Kant, *The Critique*, trans. N. Kemp Smith, p. 264 (B 3303/A246).

74 Kant, *The Critique*, trans. N. Kemp Smith, p. 93 (B 75/A 51).

75 P.F. Strawson, *The Bounds of Sense: an Essay on Kant's Critique of Pure Reason*, Methuen, London, 1966, paperback edn 1975, p. 16.

76 *ATDN*, p. 66.

77 Bohr, 'Can Quantum-Mechanical Description of Physical Reality be Considered Complete?', *Phys. Rev.*, 48, 1935, p. 700.

78 See Bohr's letter of 14 January 1936 to P. Frank, NBA: BSC 19, 3, cited above p. 140.

Chapter 11
An appraisal of Bohr's philosophy of physics

1 See M. Dummett, 'What is a Theory of Meaning? (II)', *Truth and Meaning: Essays in Semantics*, ed. G. Evans & J. McDowell, Oxford University Press, Oxford, 1976, pp. 98–100.

2 See M. Dummett, *Frege: Philosophy of Language*, p. 467; *Truth and Other Enigmas*, pp. 188, 190, 216, 218, 220, 224f.

3 M. Dummett, *Frege: Philosophy of Language*, pp. 467f.

4 See M. Dummett, 'What is a Theory of Meaning? (II)', p. 80.

5 See M. Dummett, 'What is a Theory of Meaning? (II)', p. 81.

6 See H. Høffding, *Outlines of Psychology*, trans. M.E. Lowndes, Macmillan, New York, 1896, pp. 64, 67ff. The double aspect theory was widely held throughout the nineteenth century, e.g. by F.E. Beneke, J.R. Fries, G.W.F. Hegel, G.T. Fechner, F.A. Lange, G.H. Lewes.

7 See *APHK*, p. 62.

8 S. Alexander, *Space, Time and Deity*, 2 Vols., Macmillan, London, 1920.

9 See *Essays*, p. 49.

10 See E. Schrödinger, 'Energieaustausch nach der Wellenmechanik', *Ann Phys.*, 83, 956–68, 1927; 'Are There Quantum Jumps?', *Brit. J. Phil. Sci.*, 3, 109–23, 233–42, 1952; see also M. Jammer, *The Philosophy of Quantum Mechanics*, pp. 24–33.

11 See W. Duane, 'The Transfer in Quanta of Radiation Momentum to Matter', *Proc. Nat. Acad. Sci.*, **9**, 158–64, 1923.

12 See P.S. Epstein & P. Ehrenfest, 'The Quantum Theory of the Fraunhofer Diffraction', *Proc. Nat. Acad. Sci.*, **10**, 133–39, 1924.

13 See L. de Broglie, 'Sur la possibilité de relier les phénomènes d'interférences et diffraction à la théorie des quanta de lumière', *Compt. Rend. Acad. Sci.*, **183**, 447–48, 1926; see also M. Jammer, *The Philosophy of Quantum Mechanics*, pp. 110–14, 279–91; E. MacKinnon, 'De Broglie's Thesis: A Critical Perspective', *American Journal of Physics*, **44**, 1047–55, 1976; D. Bohm, 'A Suggested Interpretation of the Quantum Theory in terms of "Hidden Variables"', *Phys. Rev.*, **85**, 166–79, 180–93, 1952.

14 See D. Bohm, 'A Suggested Interpretation', pp. 171, 174. For an enthusiastic discussion of the de Broglie–Bohm approach, see F. Selleri, 'Gespensterfelder', *The Wave-Particle Dualism*, ed. S. Diner, D. Fargue, G. Lochak & F. Selleri, Reidel, Dordrecht, 1984, pp. 101–28.

15 See R.D. Prosser, 'Quantum Theory and the Nature of Interference', *Internat. J. Th. Phys.*, **15**, 181–93, 1976. See also F. Selleri, 'Gespensterfelder', *The Wave-Particle Dualism*, pp. 124f.

16 See H. Everett, '"Relative State" Formulation of Quantum Mechanics', *Revs. Mod. Phys.*, **29**, 454–62, 1957; see also M. Jammer, *The Philosophy of Quantum Mechanics*, pp. 507ff.

17 See H. Putnam, 'Is Logic Empirical?', *Boston Studies in the Philosophy of Science*, ed. R.S. Cohen & M. Wartofsky, Reidel, Dordrecht, Vol. 5, 1968, pp. 216–41; also in H. Putnam, *Mathematics, Matter and Method: Philosophical Papers*, Cambridge University Press, Cambridge, Vol. 1, 1975, pp. 174–97 (esp. pp. 216, 174 respectively).

18 See H. Putnam, 'Is Logic Empirical?', p. 233 (*Philosophical Papers*, Vol. 1, pp. 189f).

19 See J. Bub, *The Interpretation of Quantum Mechanics*, Reidel, Dordrecht, 1974, pp. viii f., 142f.

20 See H. Putnam, *Realism and Reason: Philosophical Papers*, Cambridge University Press, Cambridge, Vol. 3, 1983, p. 266.

21 See A. Stairs, 'Quantum Logic, Realism and Value Definiteness', *Phil. Sci.*, **50**, 1983, pp. 592f.

22 See A. Stairs, 'Quantum Logic', p. 593.

23 See A. Stairs, 'Quantum Logic', pp. 598f.

INDEX